BADASS BRICKS

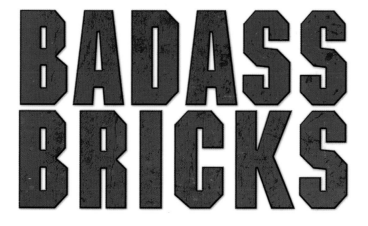

BADASS BRICKS

THIRTY-FIVE WEAPONS OF MASS CONSTRUCTION

JAKE MACKAY

SKYHORSE PUBLISHING

Skyhorse Publishing books may be purchased in bulk at special discounts for sales promotion, corporate gifts, fund-raising, or educational purposes. Special editions can also be created to specifications. For details, contact the Special Sales Department, Skyhorse Publishing, 307 West 36th Street, 11th Floor, New York, NY 10018 or info@skyhorsepublishing.com.

Skyhorse® and Skyhorse Publishing® are registered trademarks of Skyhorse Publishing, Inc.®, a Delaware corporation.

Visit our website at www.skyhorsepublishing.com.

10 9 8 7 6 5 4 3 2 1

Interior layout and graphics by Nicholas Grant

Library of Congress Cataloging-in-Publication Data is available on file.

ISBN: 978-1-62636-304-5

Printed in China

CHAPTER THREE
HEAVY ASSAULT

CHAPTER FOUR
SPECIAL WEAPONS

INTRODUCTION

This LEGO instructional manual is an awesome, badass compilation of some of history's most destructive weapons. Included inside these pages are medieval weapons, such as the claymore sword, the war hammer, the battering ram, and the siege tower, and more modern weapons of mass destruction from the twentieth century, such as the Paris gun, the V-2 missile, the VIPeR combat robot, and the crossbow pistol.

Considering the extreme power behind these weapons and the determination of those who wielded them, the weapons in this manual are some of the most well known throughout history. The rack, a medieval torture device used for interrogation with the intention of pulling its victim limb from limb, was used throughout the Middle Ages. The Landkreuzer P. 1000 battle tank would have been larger than any military device ever built and would have weighed over one thousand tons. The M1114 Humvee was considered so awesome that a civilian version was produced for sale. No matter which weapon you choose to build, there is something awe-inspiring to discover.

This manual is divided into straightforward sections (hand-to-hand combat, ranged weapons, heavy assault, and special weapons), and each weapon is comprehensively built, piece by piece. A few of the weapons are fully functional, like the crossbow pistol, the rack, and the Landkreuzer, which can be hooked up to a battery-powered remote. The full-scale models, like the claymore sword, which is four feet in length, bring a realistic aspect to these historical weapons. At the end of this book is a Bill of Materials for each project, making it easy to see which pieces are needed for each model.

CHAPTER ONE
HAND-TO-HAND COMBAT

WAR HAMMER
SCALE: 1:20

A fourteenth- to fifteenth-century medieval weapon, war hammers have long shafts with an iron head that was used to strike opponents. In medieval wartime, the blunt end could not penetrate armor, but it could do significant crushing damage or create such force against an enemy's helmet that it caused a concussion in the wearer. Over time, armor was produced in part with steel, which was so strong that battle-axes and swords would simply ricochet upon impact. As a result, the spiked end of the war hammer was developed to pierce weak spots in armor and impale the enemy. While most depictions are very similar to modern hammers, other examples are the squared-off, mallet-like hammer used by Thor or trolls in popular lore.

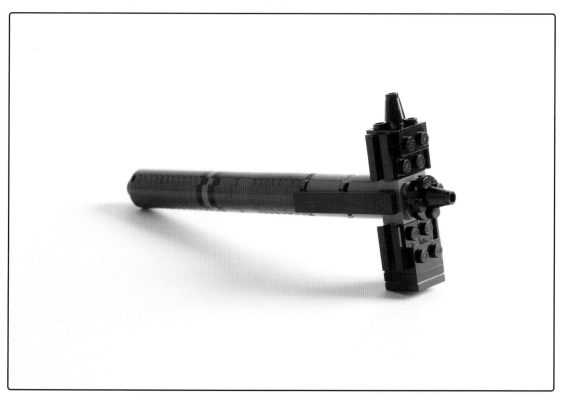

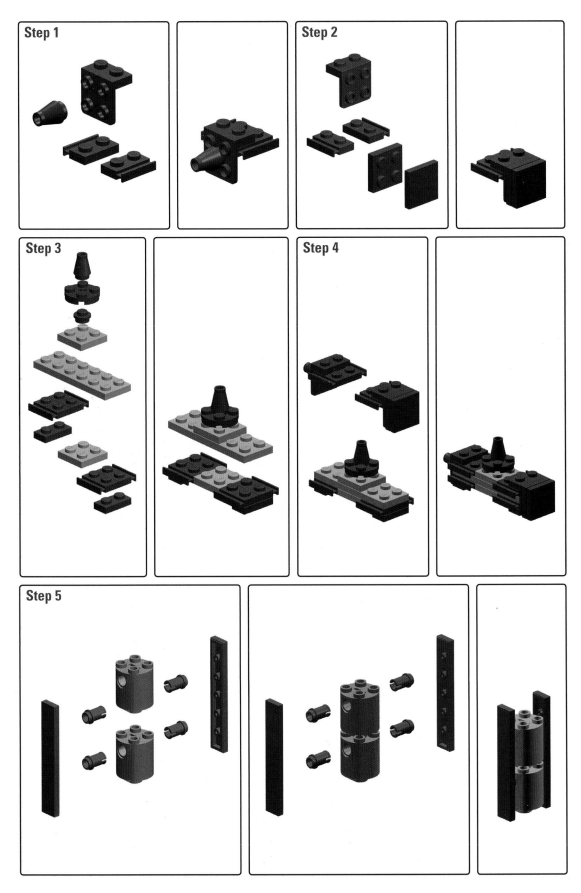

Step 1

Step 2

Step 3

Step 4

Step 5

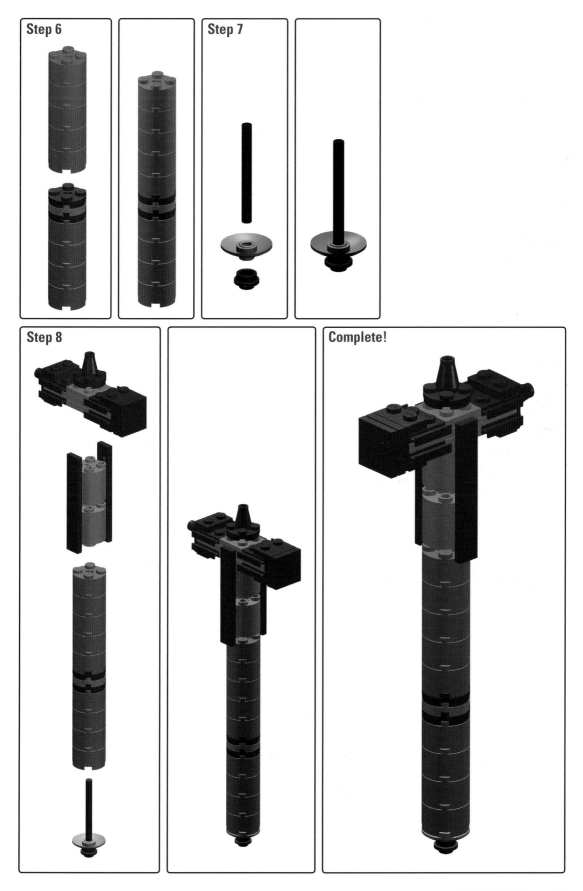

Step 6

Step 7

Step 8

Complete!

KATANA

SCALE: 1:1

The ultimate weapon of the samurai,
the *katana* is a slender, single-edged
sword that originated in Feudal Japan.
Known for its strength and saber-
sharp blade, the samurai sword makes
its competitors look like butter knives.

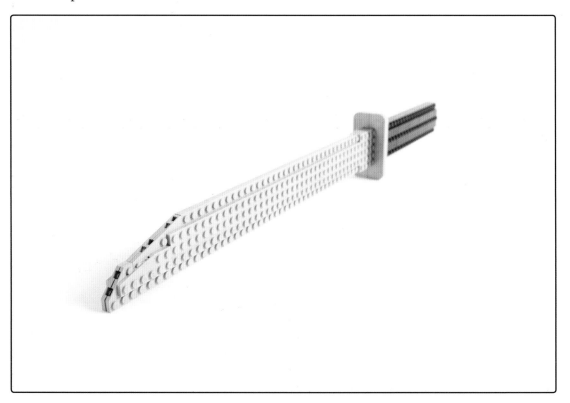

FIRST HALF OF THE KATANA

Step 1

Step 2

Step 3

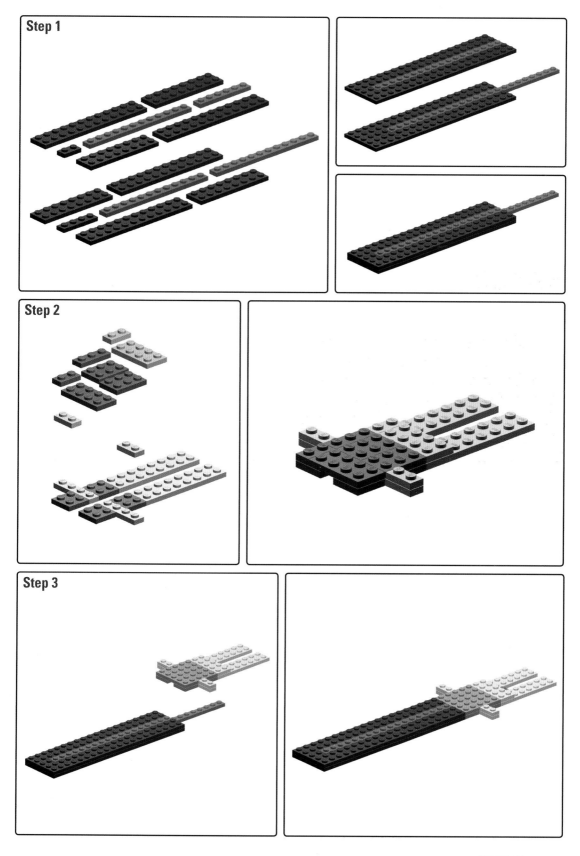

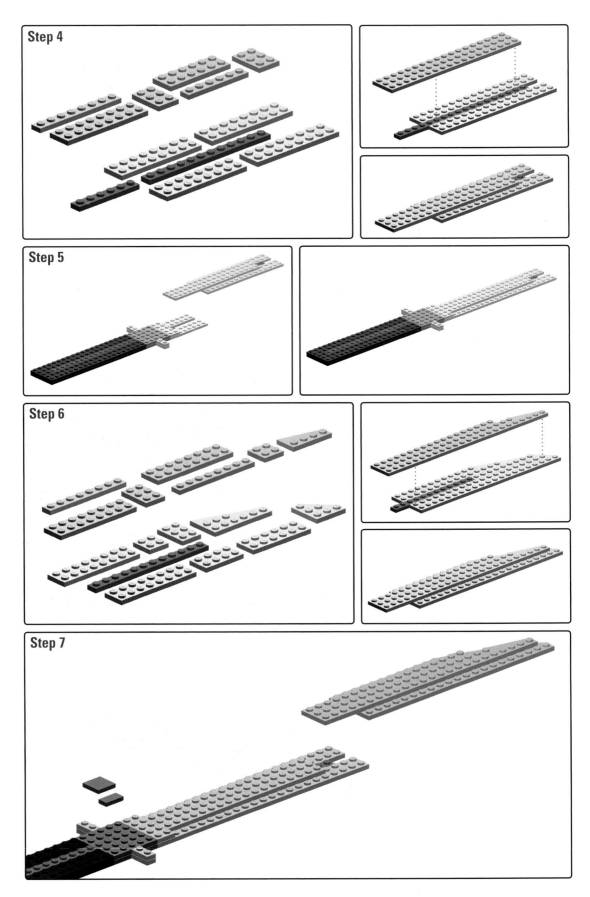

Step 4

Step 5

Step 6

Step 7

Done

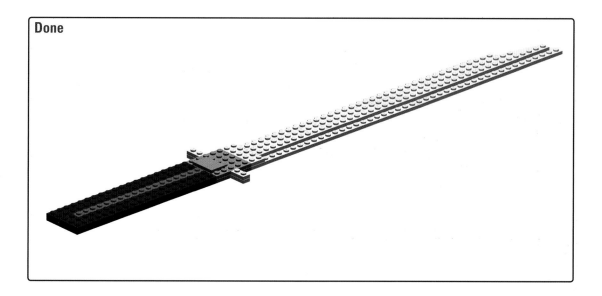

SECOND HALF OF THE KATANA

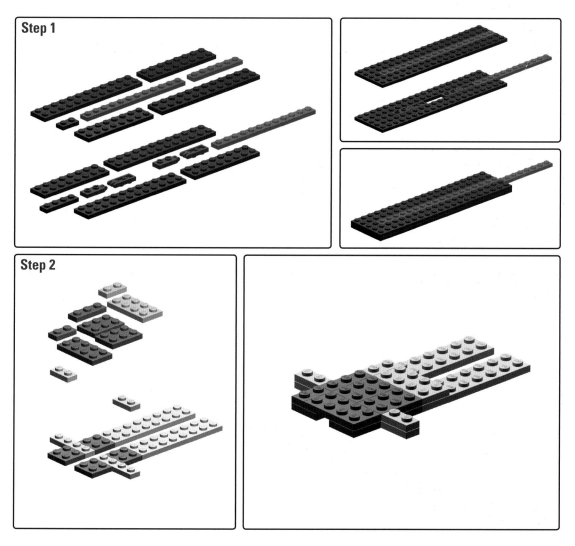

Step 1

Step 2

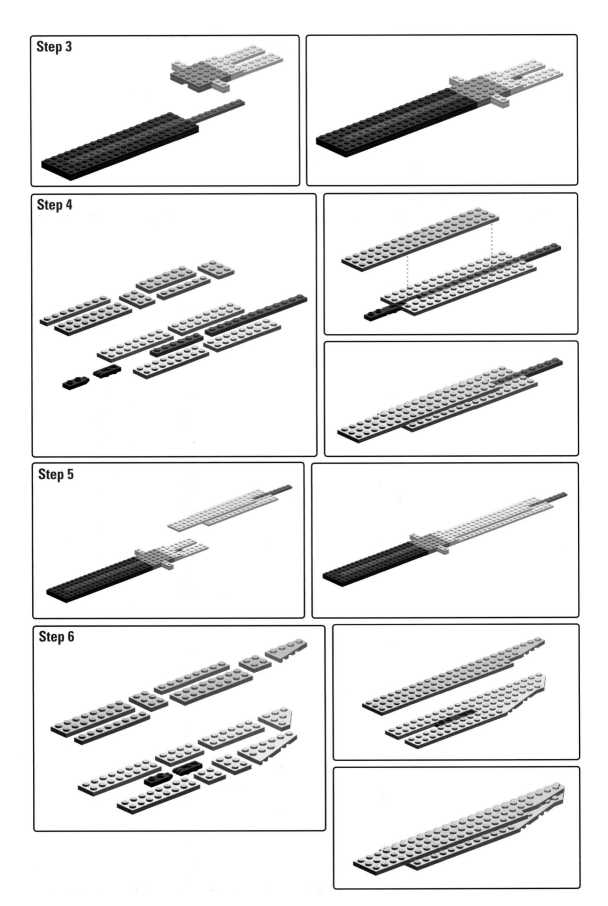

Step 3

Step 4

Step 5

Step 6

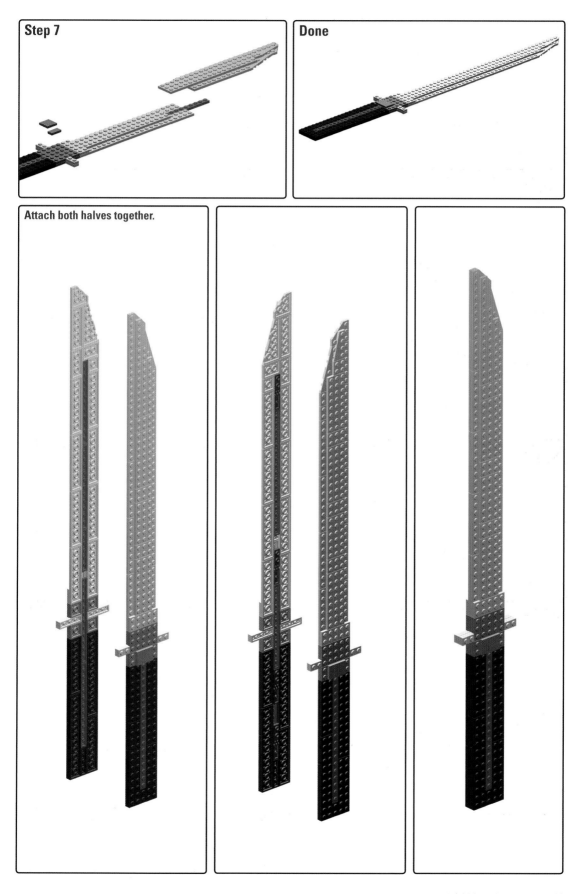

Step 7

Done

Attach both halves together.

FINAL ATTACHMENTS

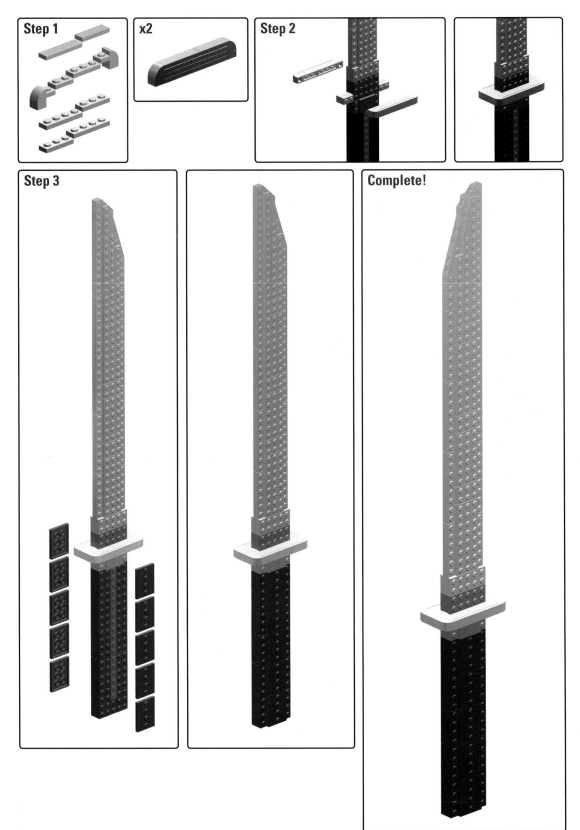

Step 1

x2

Step 2

Step 3

Complete!

MAQUAHUITL
SCALE: 1:1

Not your fraternity's pledge paddle.
This Pre-Aztec hand-to-hand wooden
club was traditionally lined with
teeth-like prismatic blades made of
obsidian. Sometimes up to four feet
long, it could be constructed with
either a one-handed or a two-handed
grip, packing a razor-sharp blow that
could decapitate opposing warriors.
While no authentic *maquahuitls* have
survived, they were thought to be used
by Aztec military forces and may have
even been wielded against sixteenth-
century Spanish *conquistadors*.

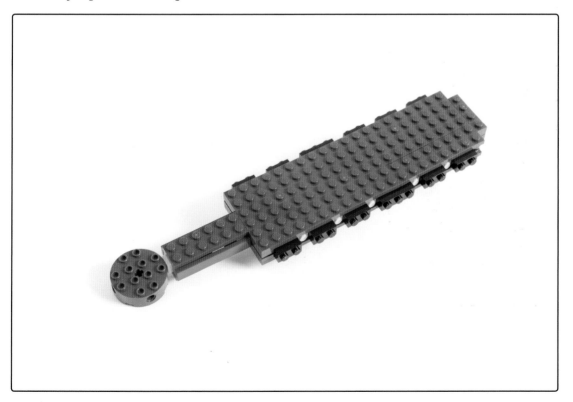

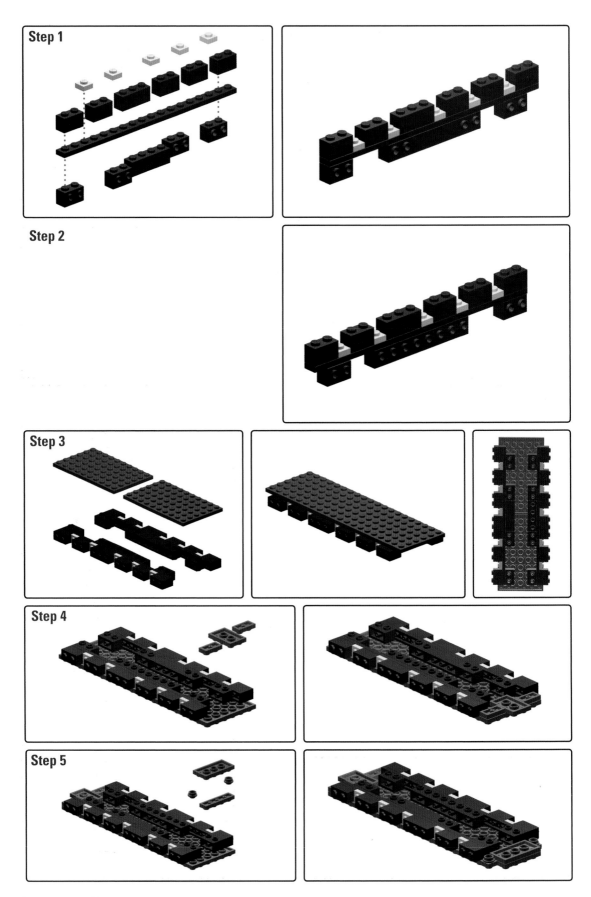

Step 1

Step 2

Step 3

Step 4

Step 5

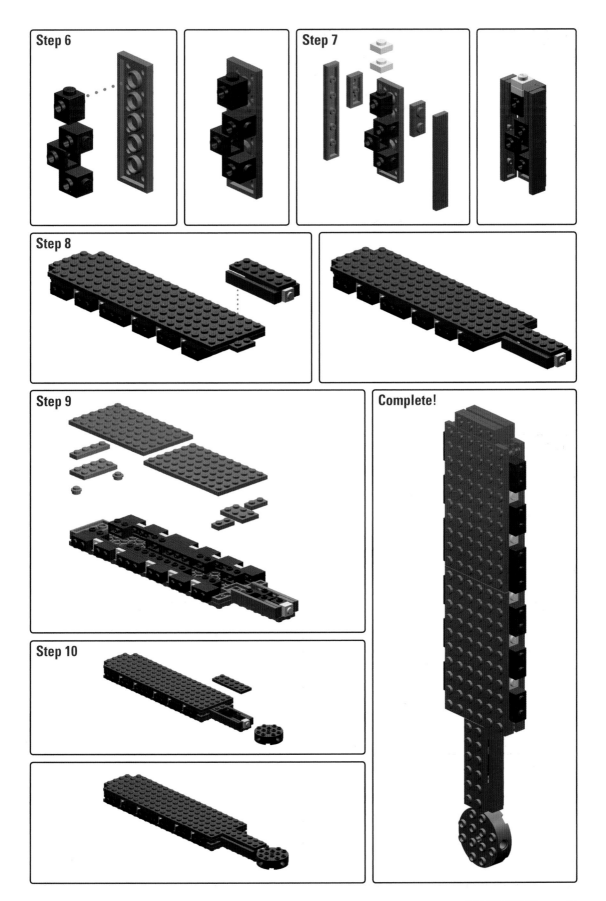

Step 6

Step 7

Step 8

Step 9

Step 10

Complete!

MISERICORDE
SCALE: 1:1

With a name sounding like "misery," the twelfth-century *misericorde* actually derives from the Latin word for "mercy." Used as a form of euthanasia for mortally wounded knights, this needle-pointed dagger and its thin, steely blade could be thrust below the knight's neck and down to the heart for a *coup de grace*, or to put him out of his misery. When used offensively, this sharp dagger could also be used to pierce through weak points in armor or stab through open eyeholes against enemies during a grapple. Offshoots of this weapon are the fifteenth-century rondel dagger and the Italian *stiletto*.

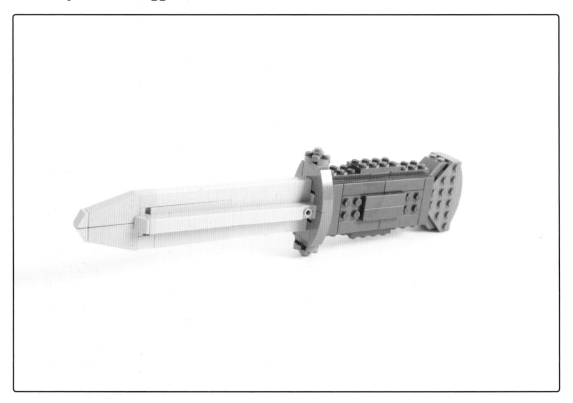

BLADE

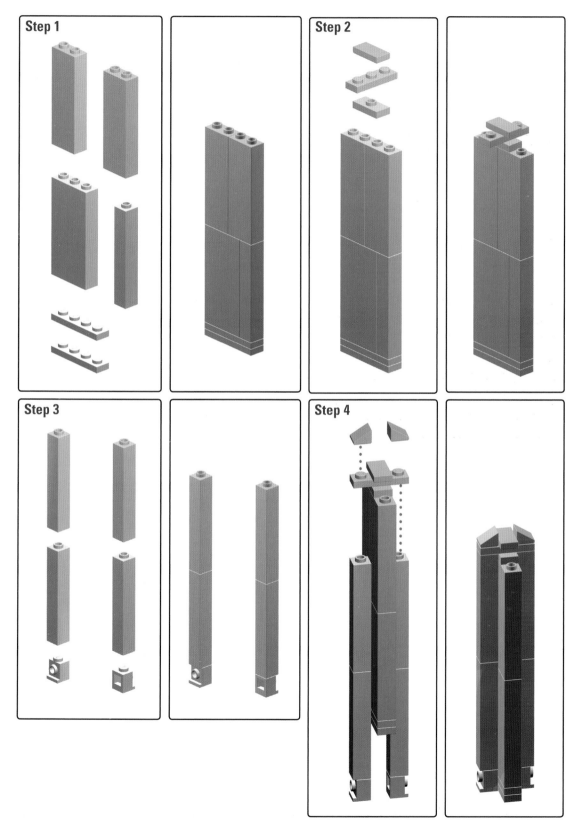

Step 5

Step 6

HANDLE CORE

Step 1

Step 2

Step 3

HANDLE SIDE COVER

Step 1

Step 2

Step 3

x2

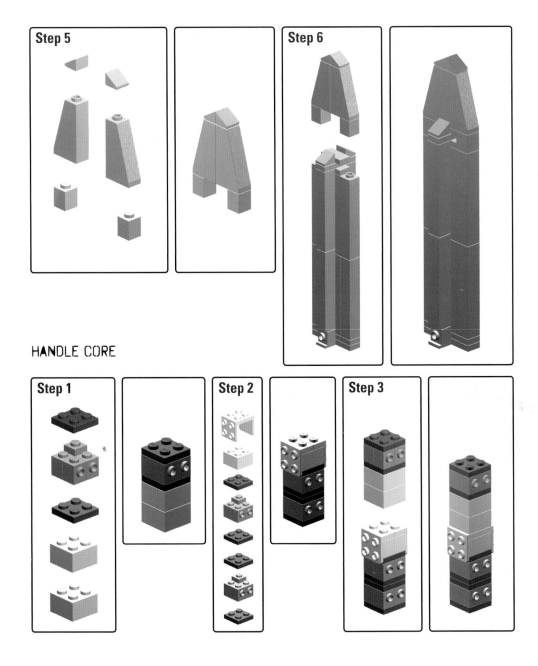

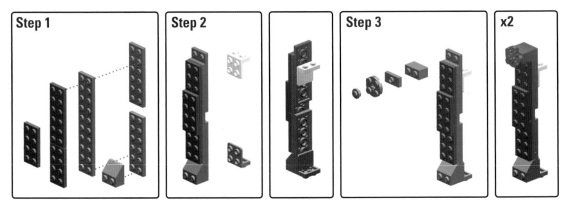

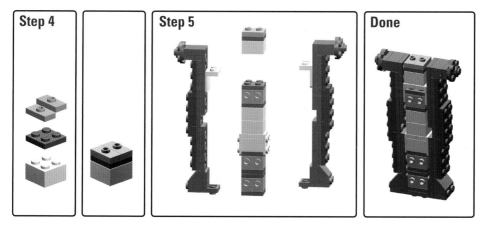

Step 4

Step 5

Done

WIDE HANDLE COVER

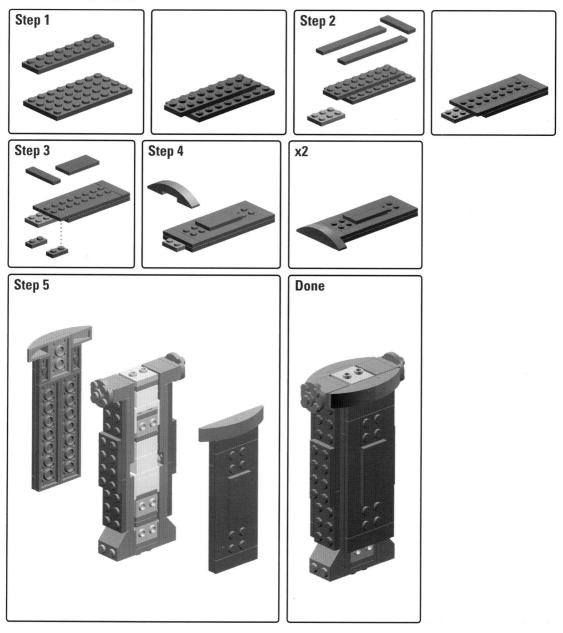

Step 1

Step 2

Step 3

Step 4

x2

Step 5

Done

FINAL ATTACHMENTS

Step 1

Step 2

Step 3

Complete!

Step 4

BATTLE-AXE
SCALE: 1:1

The battle-axe is a very early
form of weaponry: the first
was likely a tool repurposed
for battle, as is the case with
many early weapons. In its
simplest form, a battle-axe
consists of a curved blade
attached to a stick. The Viking
battle-axes were light with
a thin, sharp blade just right
for hacking and slashing.

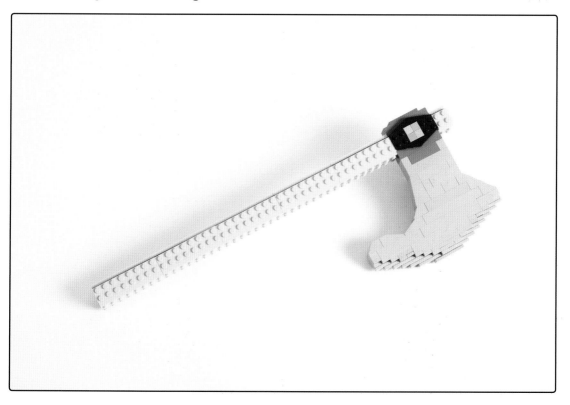

AXE HANDLE

Step 1

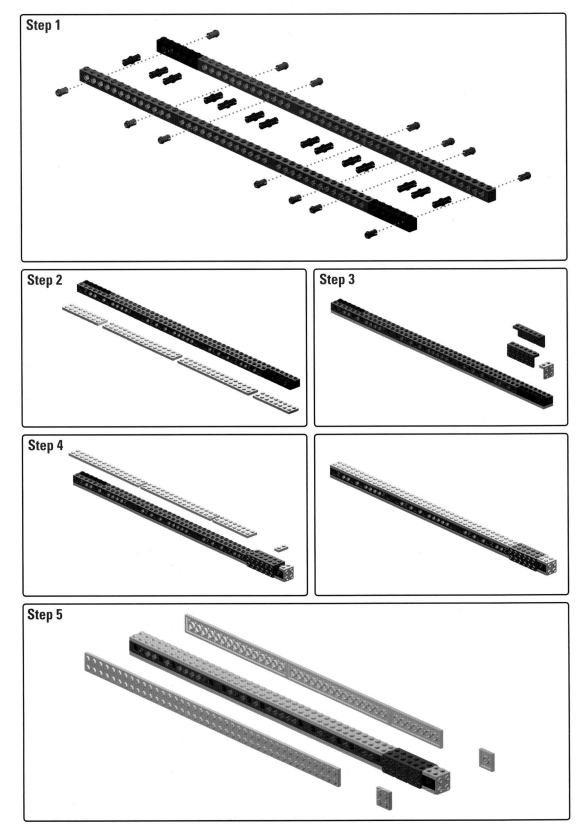

Step 2

Step 3

Step 4

Step 5

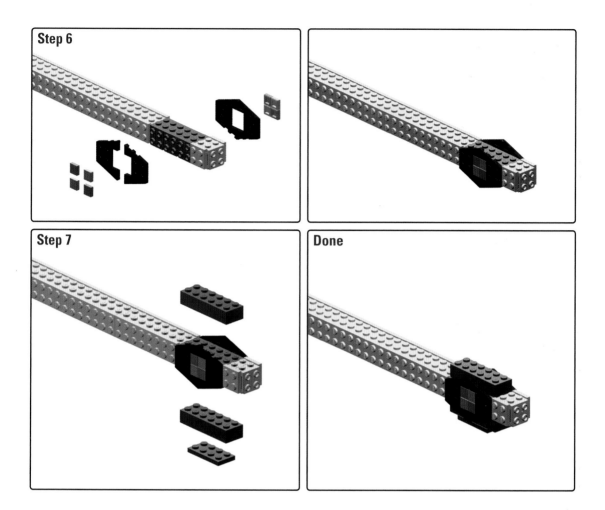

AXE BLADE

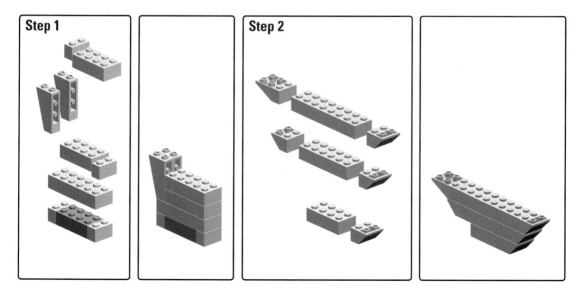

Step 3

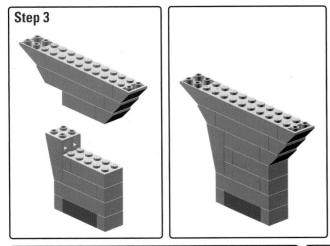

Step 4

Step 5

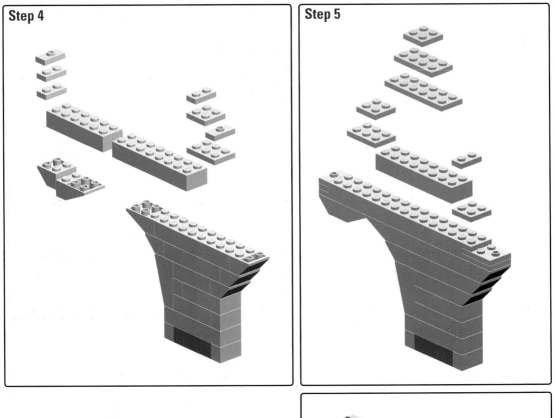

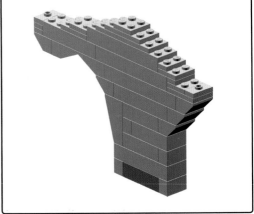

Step 6

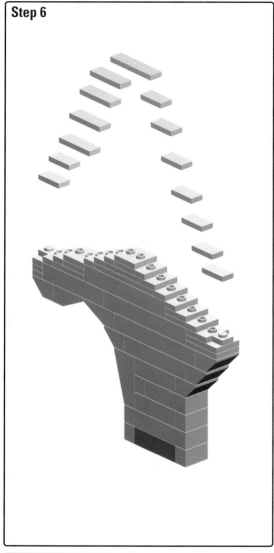

Done

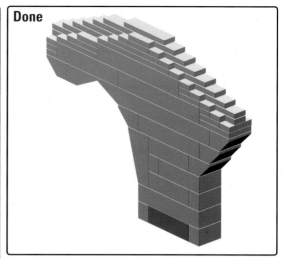

FINAL ATTACHMENTS

Step 1

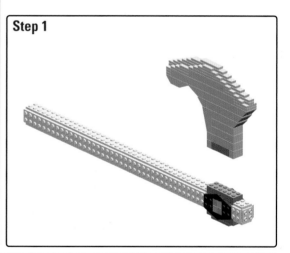

Complete!

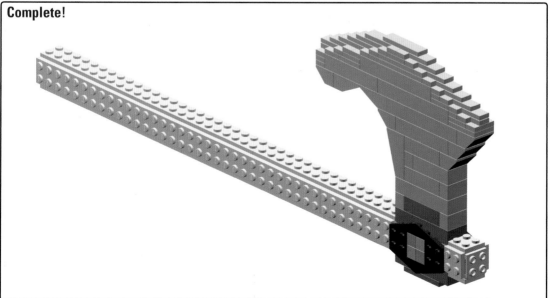

CLAYMORE
SCALE: 1:1

A Scottish great sword, the claymore was a two-handed longsword distinguished by its cross-hilt of forward-tilting quillons, the guards that separate the blade from the handle. Used from the fifteenth to seventeenth centuries, this powerful Highlander weapon was double-edged and ranged from three to nearly four feet in length. As a full-scale model, this giant of a sword has a technic sub-structure connecting the handle, cross-hilt, and blade, with the LEGO bricks' studs facing inward towards the technic beams.

BLADE

Step 1

Step 2

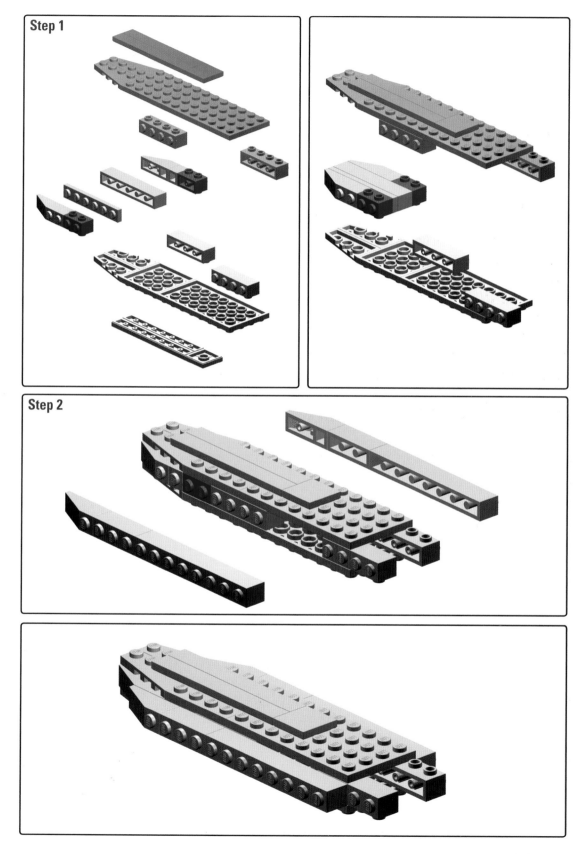

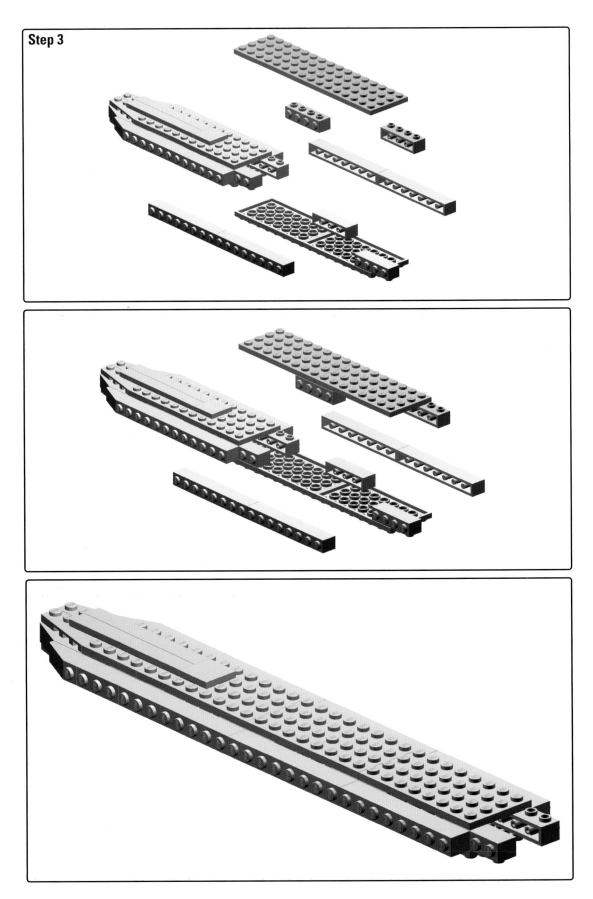

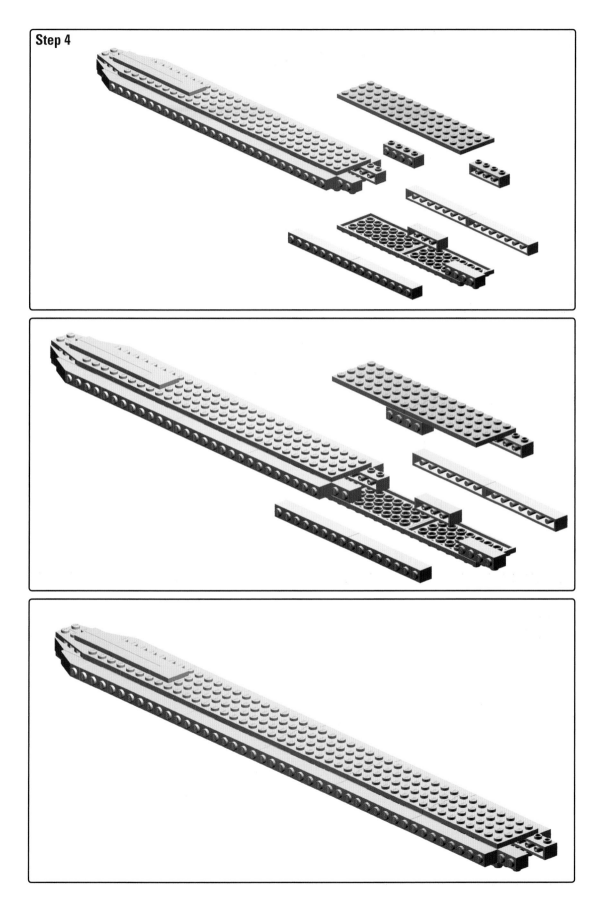

Step 5

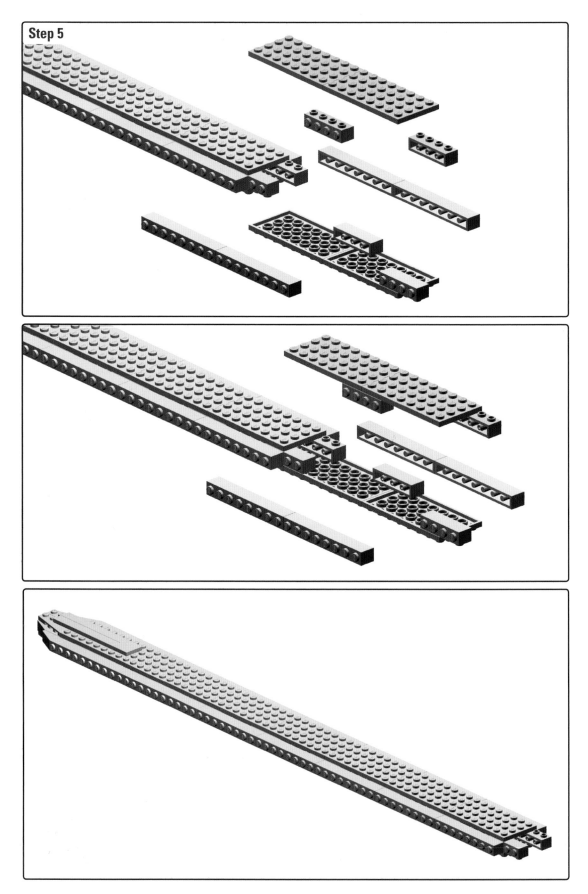

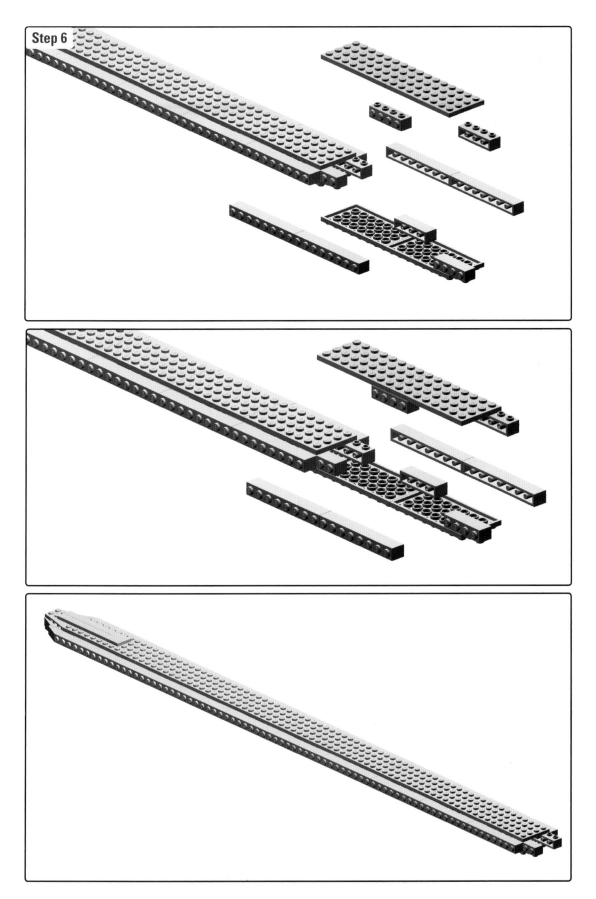

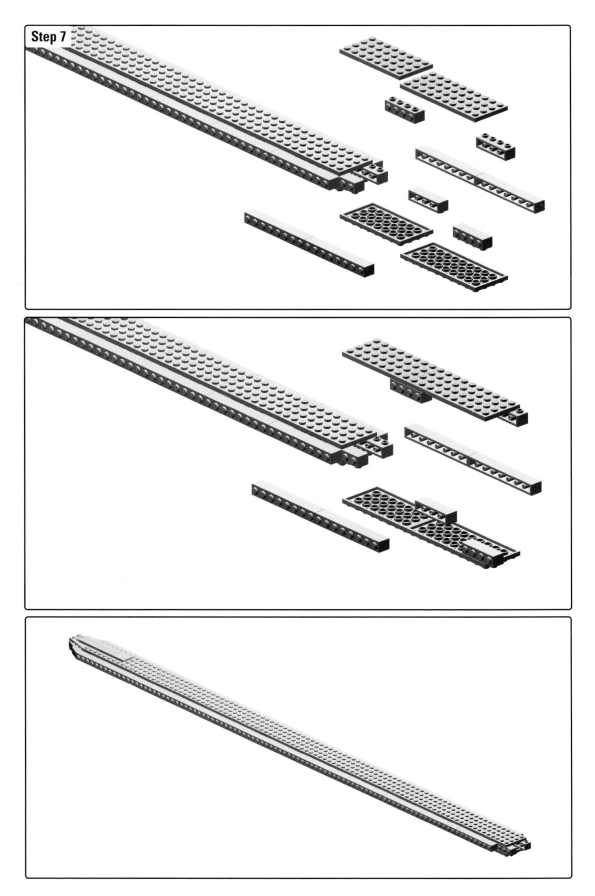

Step 7

Step 8

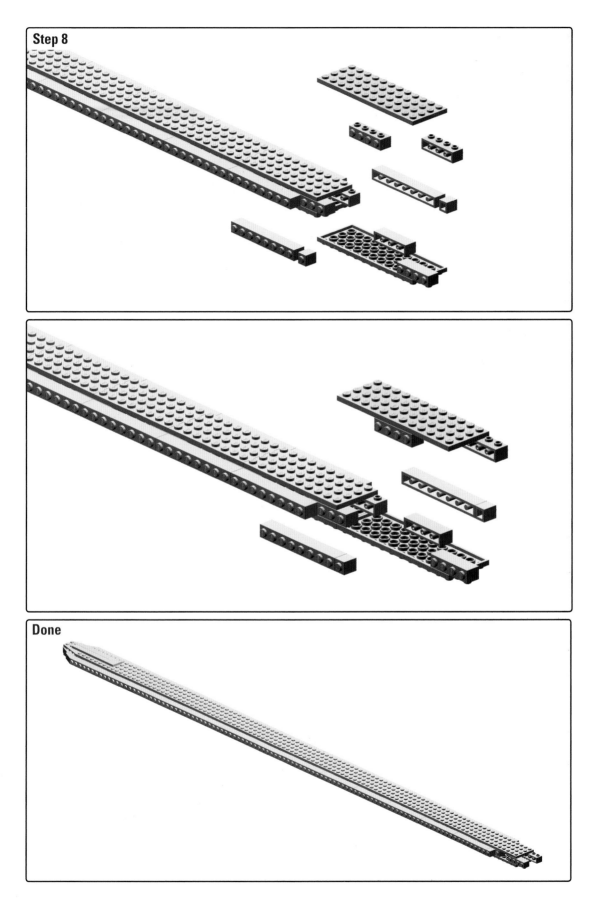

Done

HANDLE

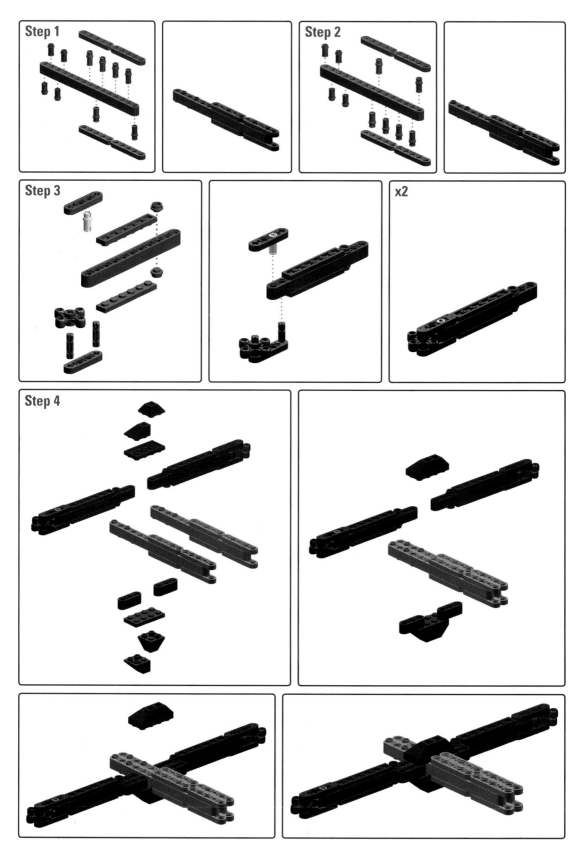

Step 1

Step 2

Step 3

x2

Step 4

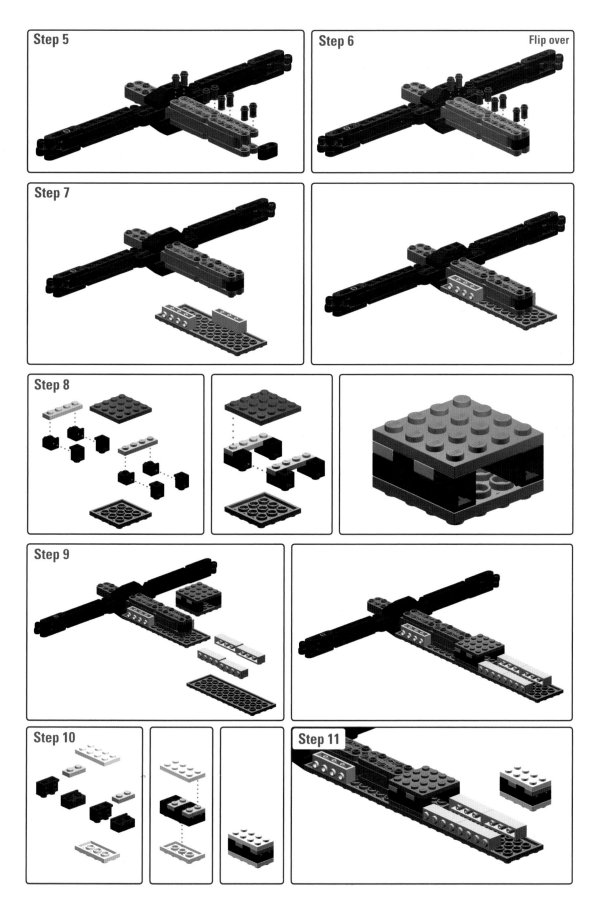

Step 5

Step 6   Flip over

Step 7

Step 8

Step 9

Step 10

Step 11

Step 12

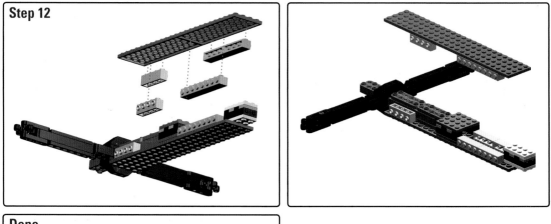

Done

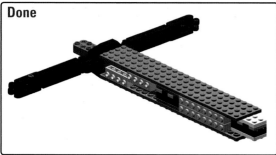

HANDLE DETAILS

Step 1 **x2** **Step 2** **x2**

Step 3 **Step 4**

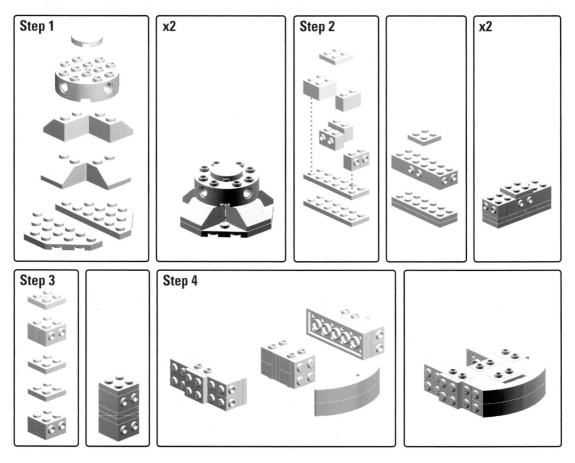

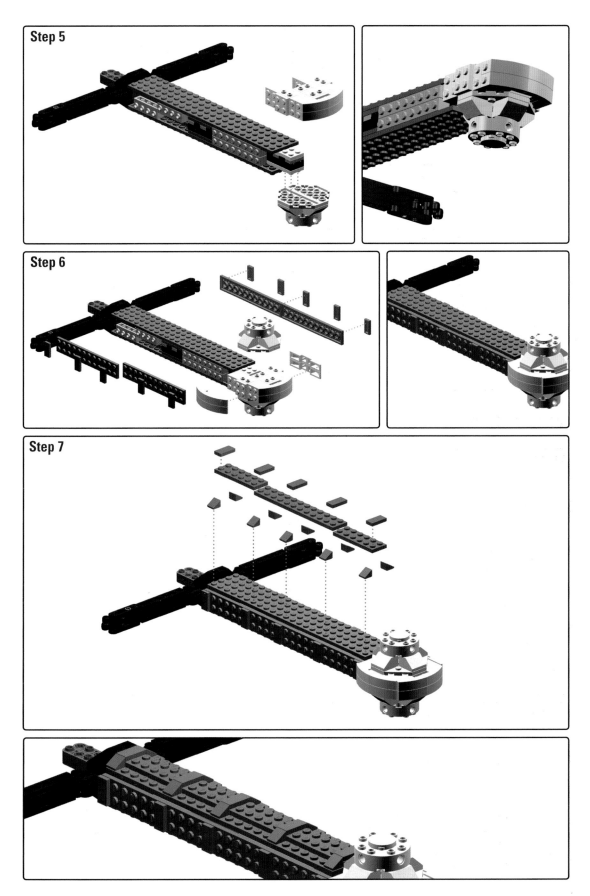

Step 5

Step 6

Step 7

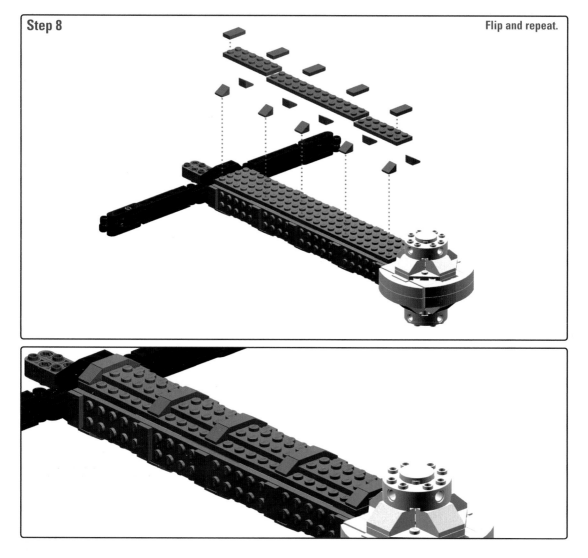

FINAL ATTACHMENTS

Step 1

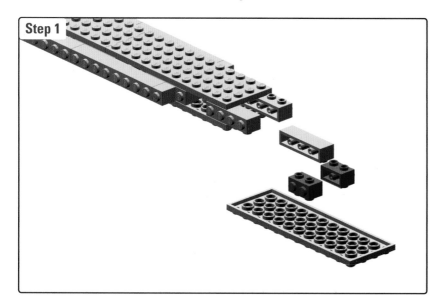

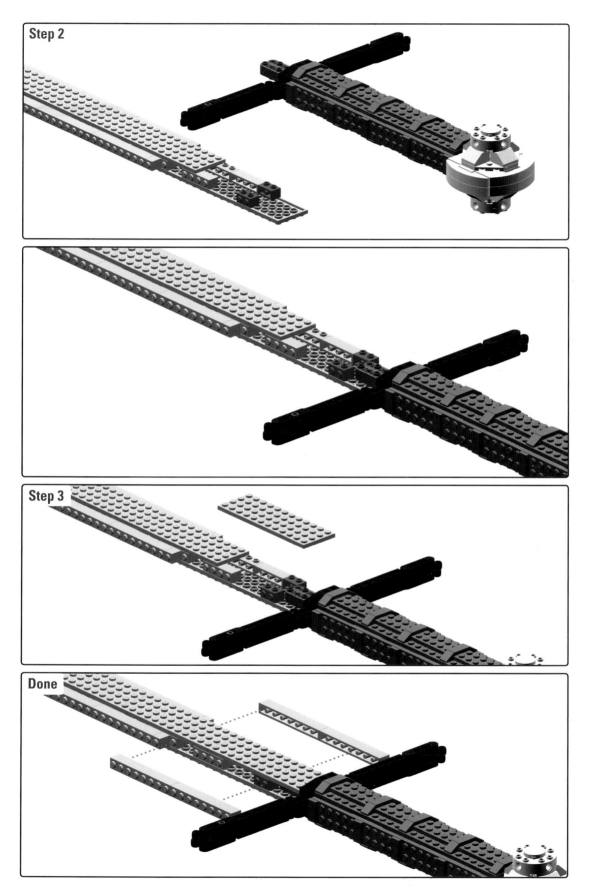

Step 2

Step 3

Done

BLADE DETAILS

Step 1

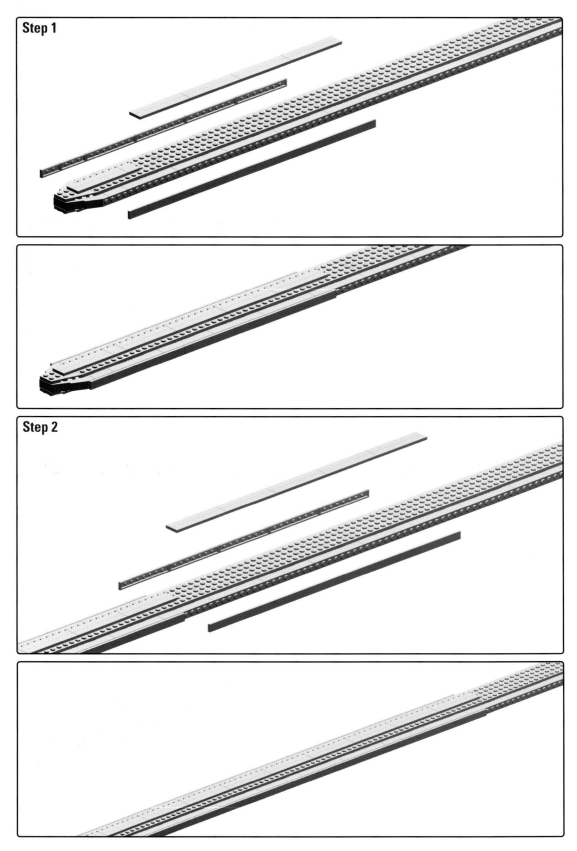

Step 2

Step 3

Done
Repeat tiling on the opposite side.

Complete!

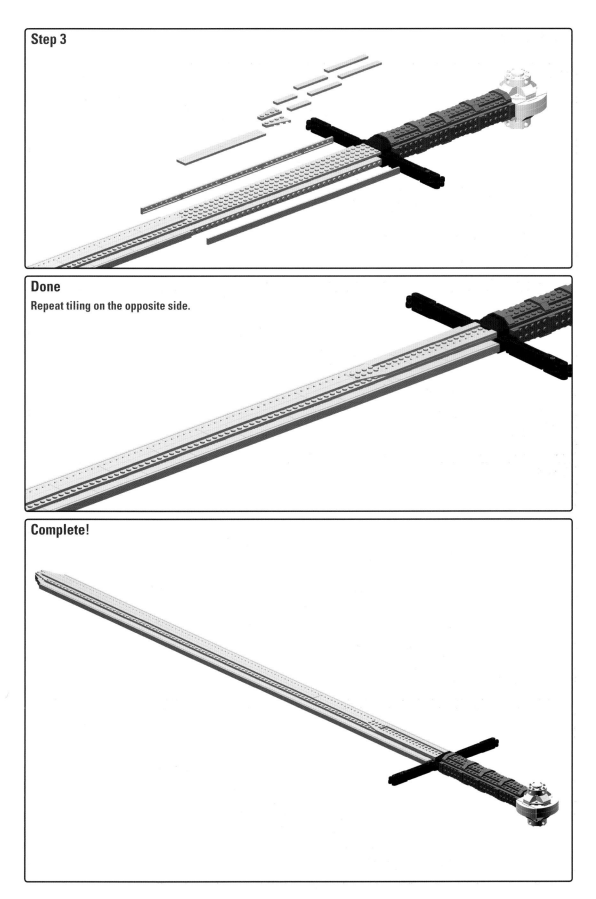

HALF-MOON AXE
SCALE: 1:20

The half-moon axe is a fantasy variant of the battle-axe. Used exclusively as an execution tool, it has an intimidating blade big enough to lop off a head in one go and a corresponding vicious spike on the opposite end. The name comes from the curved blade, which resembles a half-moon.

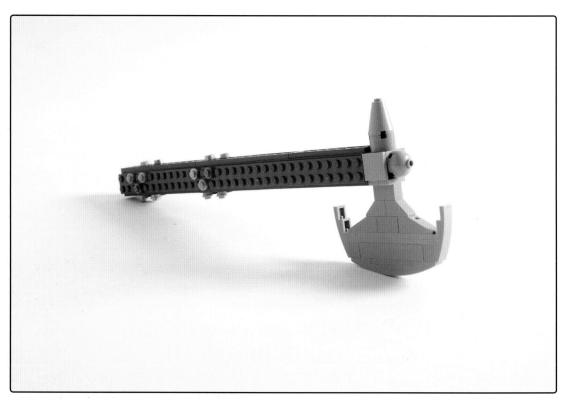

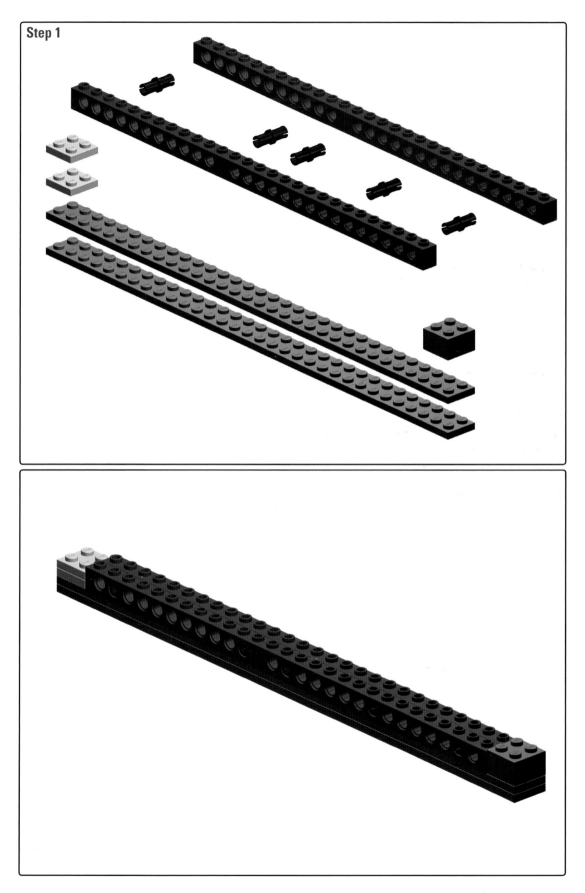

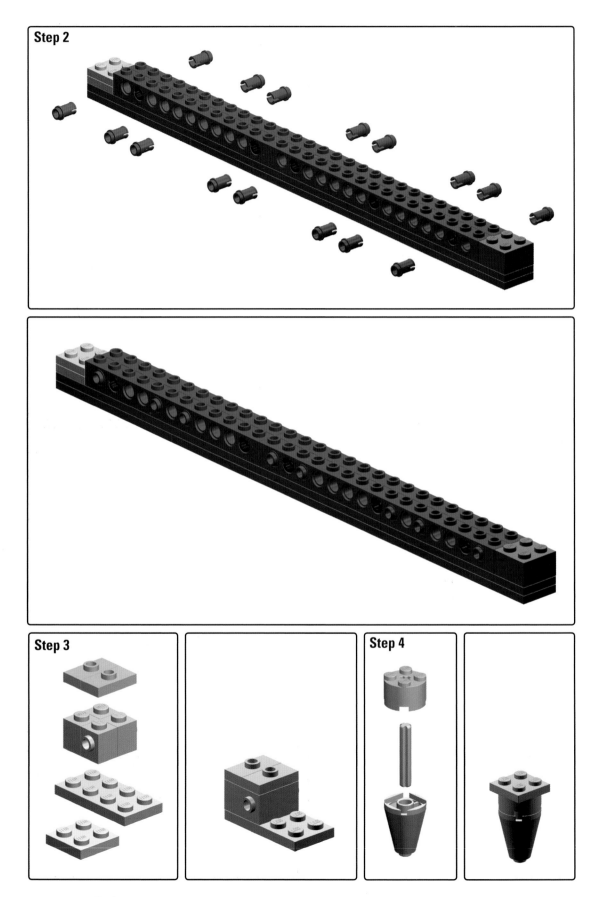

Step 2

Step 3

Step 4

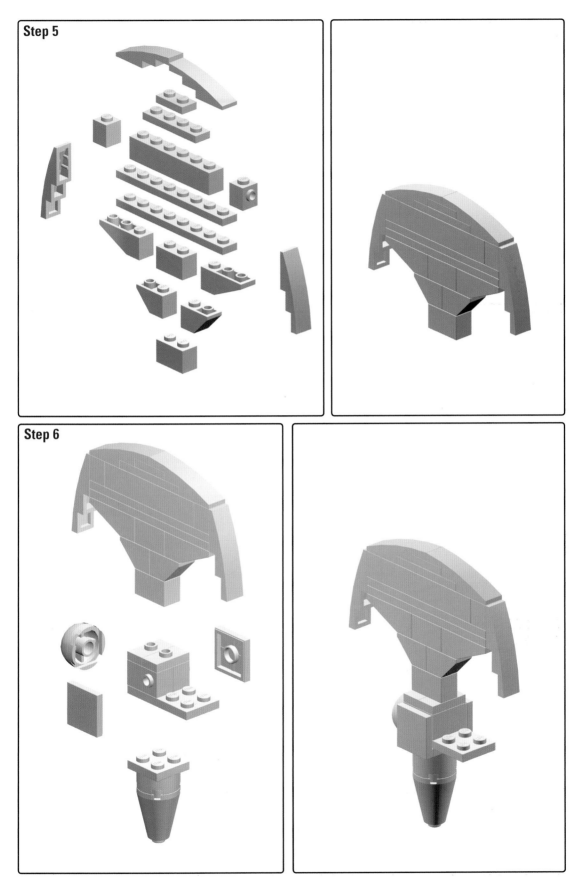

Step 5

Step 6

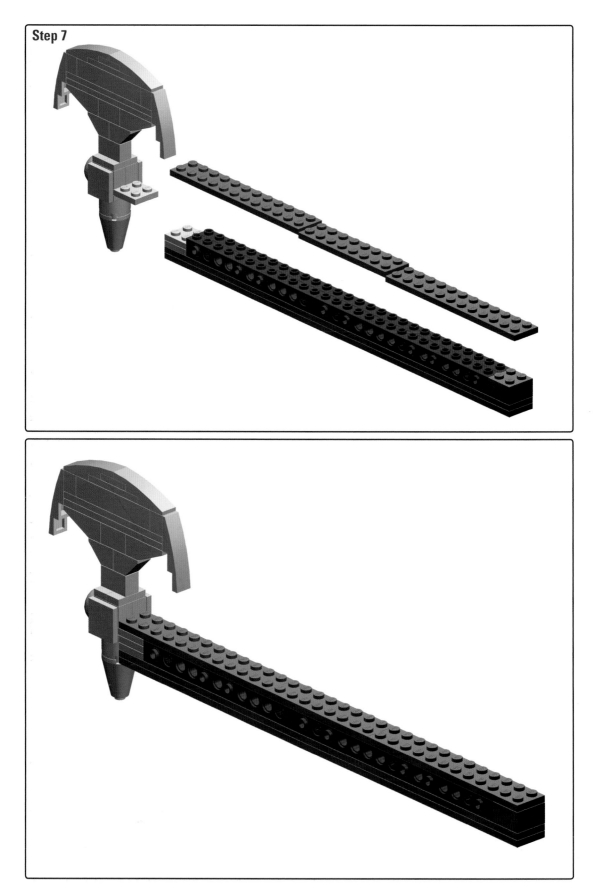

Step 7

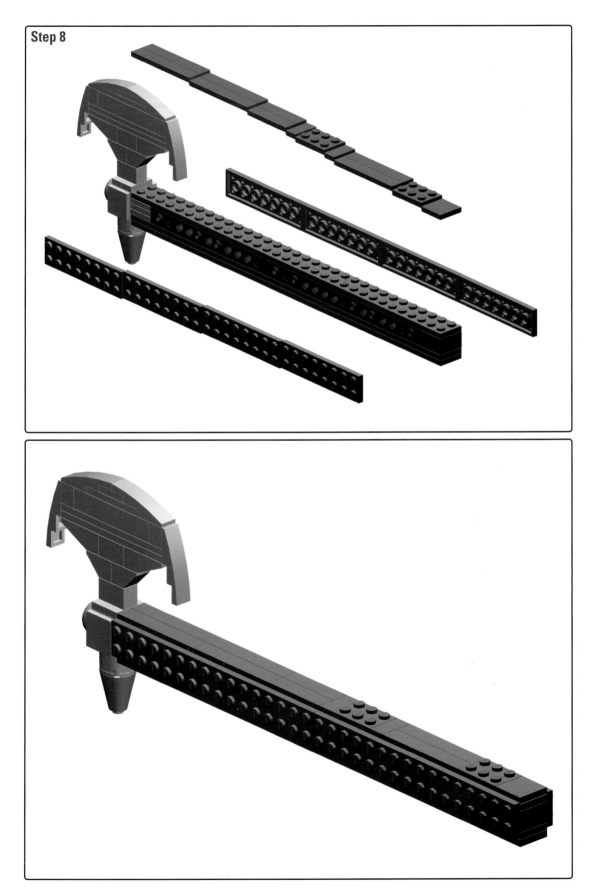

Step 8

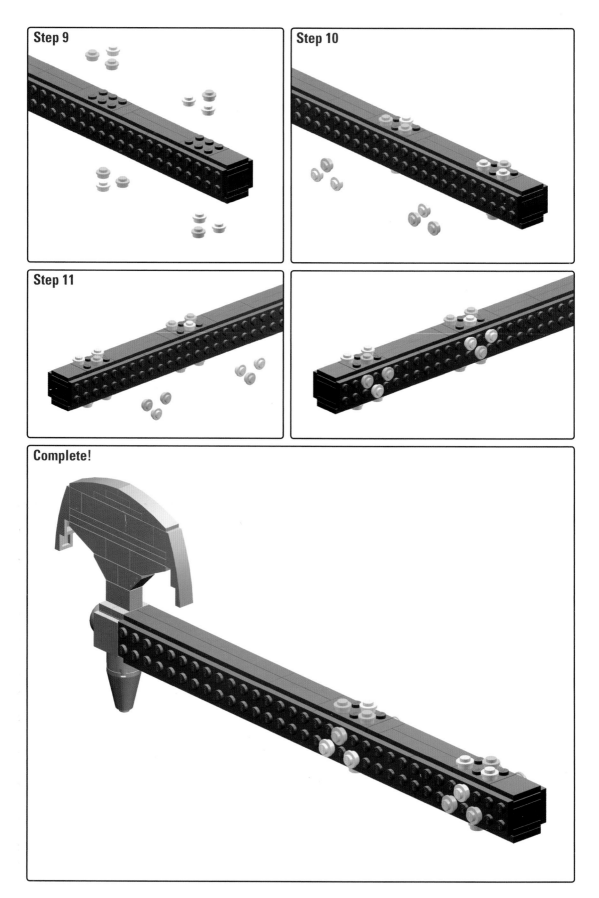

Step 9

Step 10

Step 11

Complete!

MORNING STAR

SCALE: 1:1

A devil of a weapon, the morning star is a wooden-handled club resembling a mace. Making its way onto battlefields around the fourteenth century, it was particularly widespread in Germany. The end of the shaft was a large metal club with huge spikes. The wielder could use the morning star to beat and puncture his opponent with the spikes.

Step 1

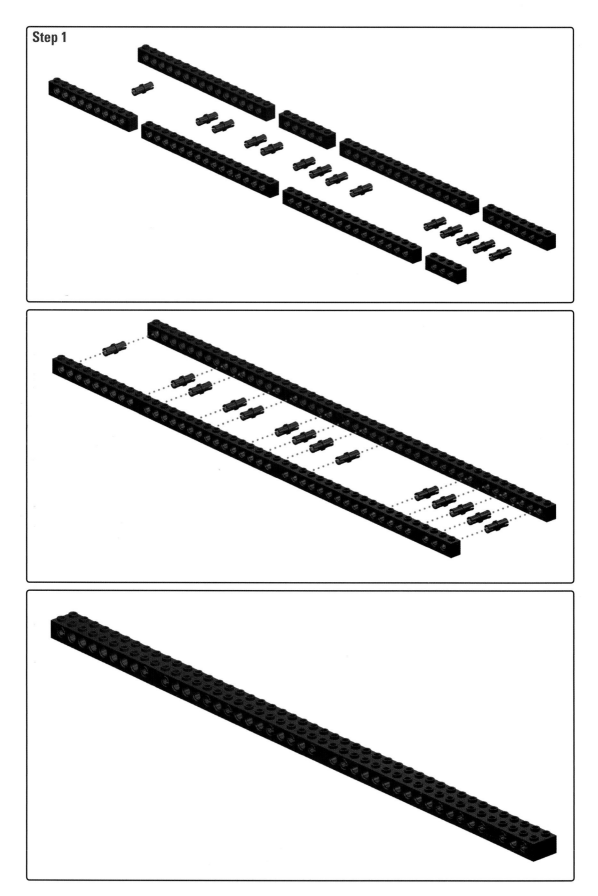

Step 2

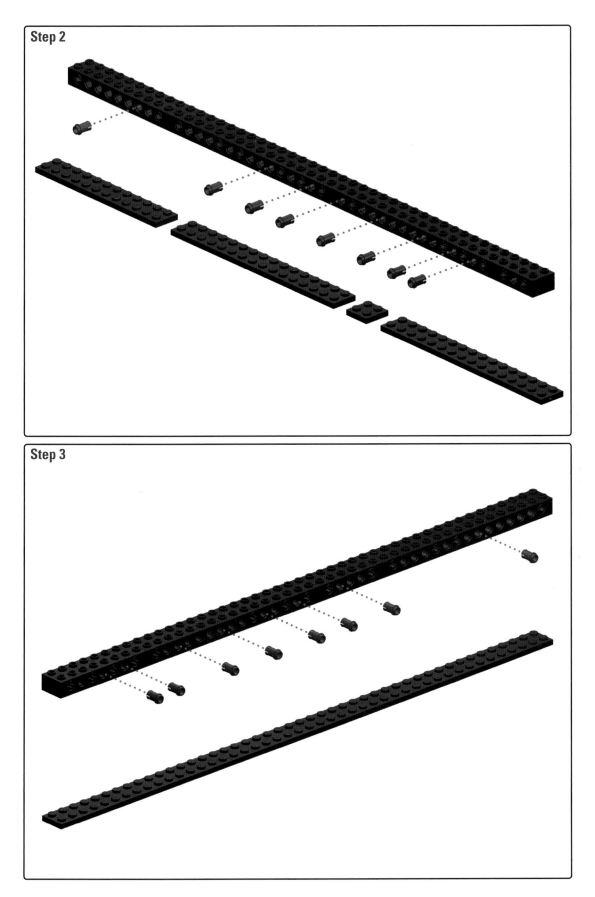

Step 3

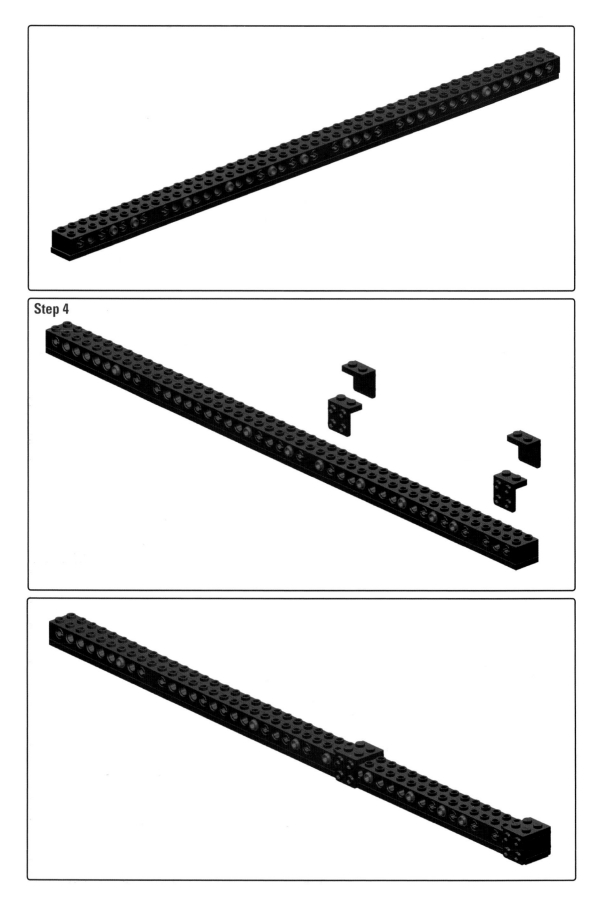

Step 4

Step 5

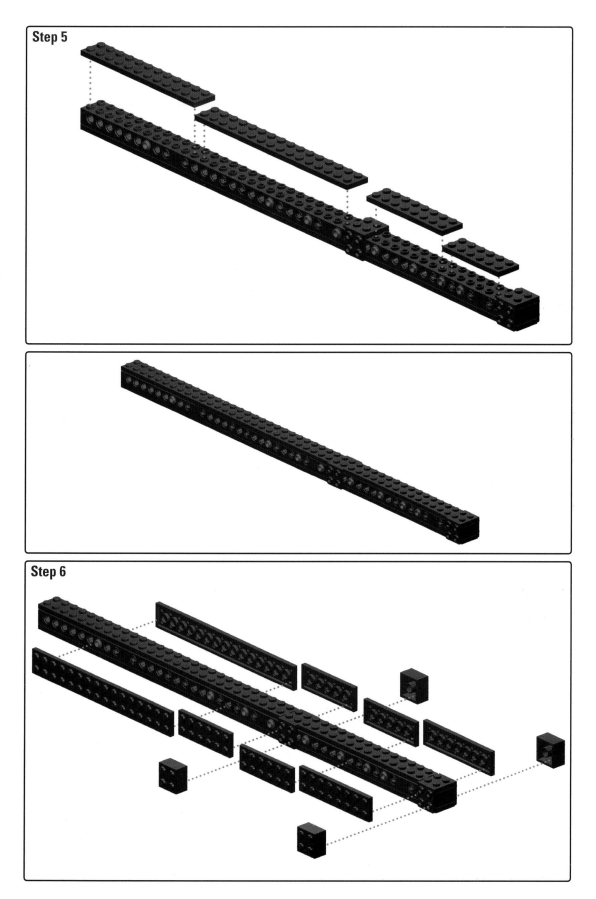

Step 6

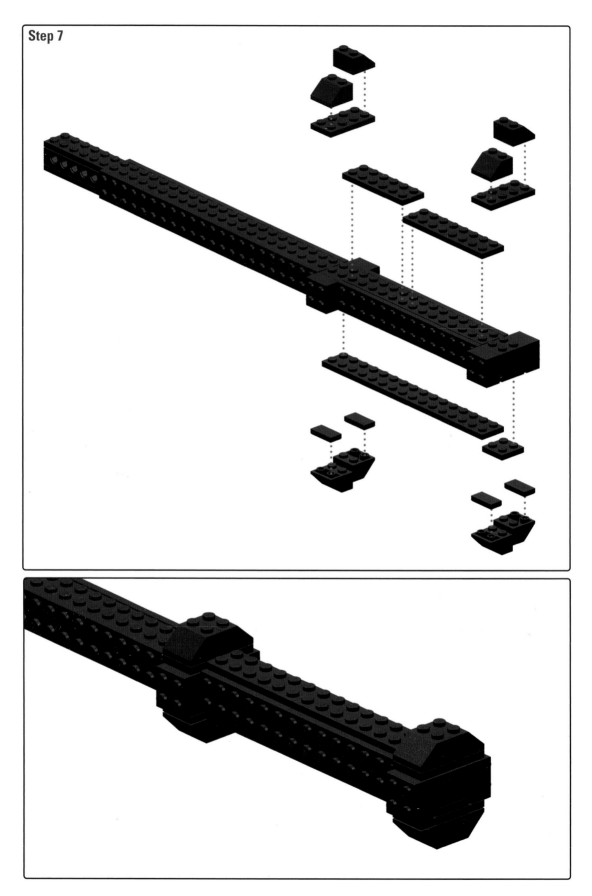

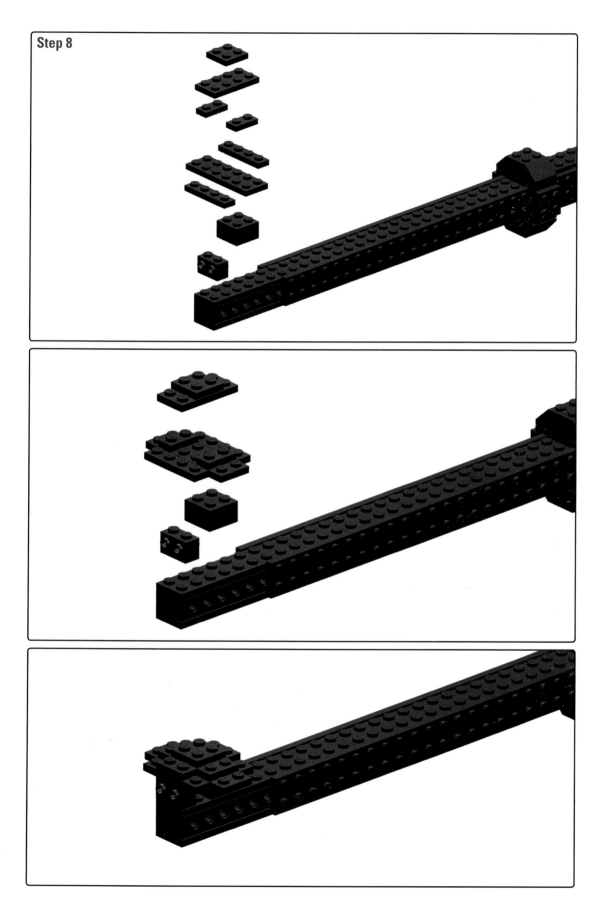

Step 8

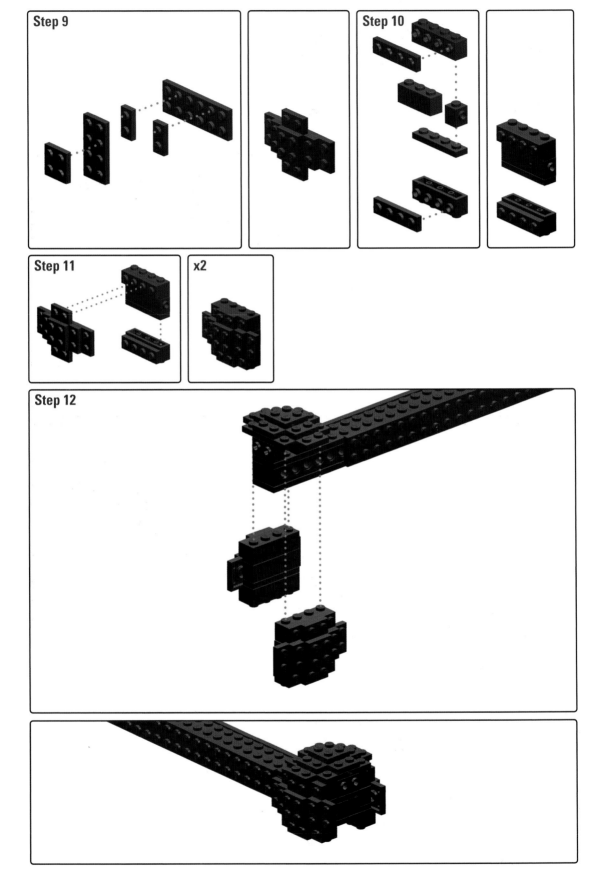

Step 9

Step 10

Step 11

x2

Step 12

Step 13

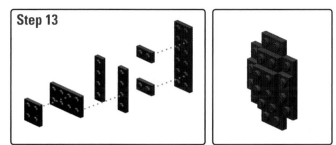

Step 14

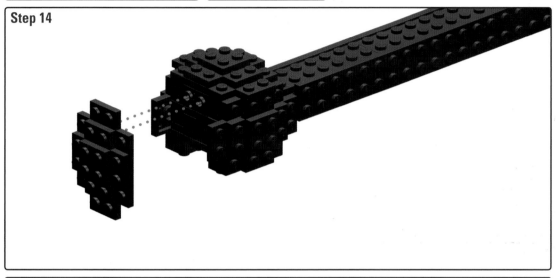

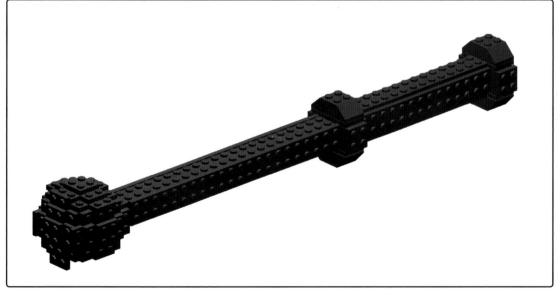

Step 15

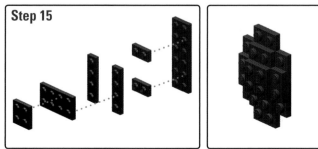

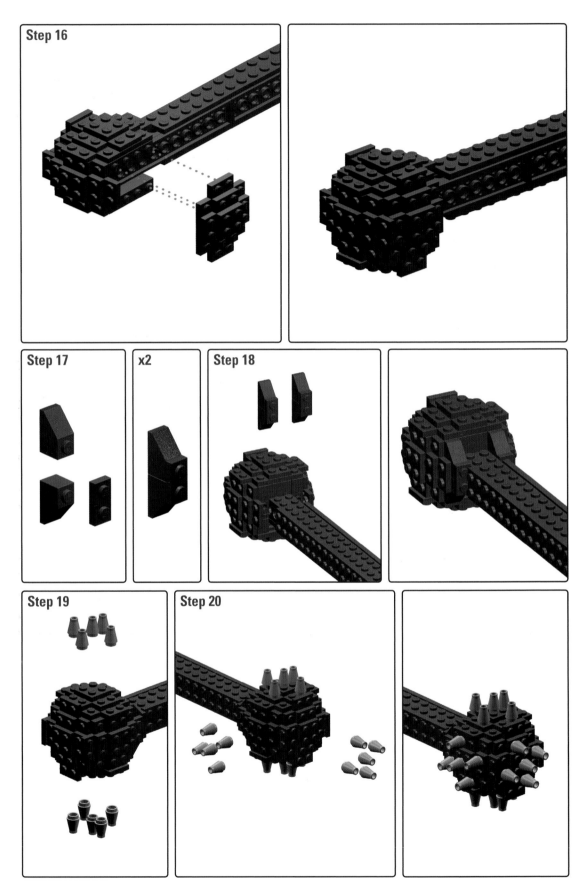

Step 16

Step 17

x2

Step 18

Step 19

Step 20

Step 21

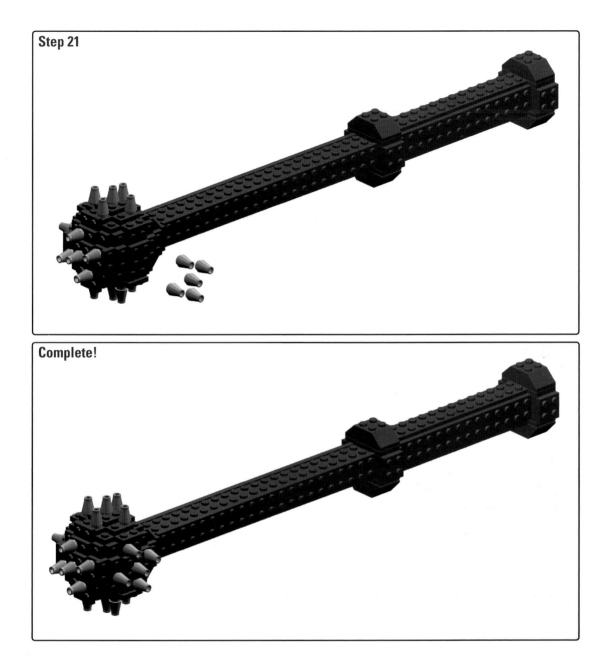

Complete!

BOWIE KNIFE
SCALE: 1:1

Bring a knife to a gun battle. Made famous by Colonel Jim Bowie during the 1827 "Sandbar Fight," Bowie was shot, stabbed, and beaten to a pulp but managed to come out victorious with the help of this beastly knife. This sheath knife has a short handle with a cross-guard and a clip-point (curved) blade that can be anywhere from five to twenty-four inches long. Popular for hunting, defense, or as an everyday tool, this knife is the American standard.

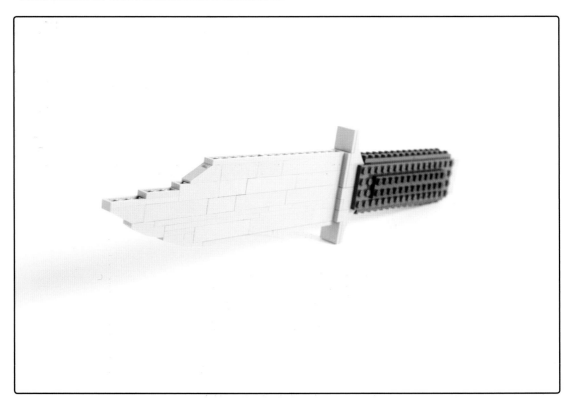

BLADE

Step 1

Step 2

Step 3

Done

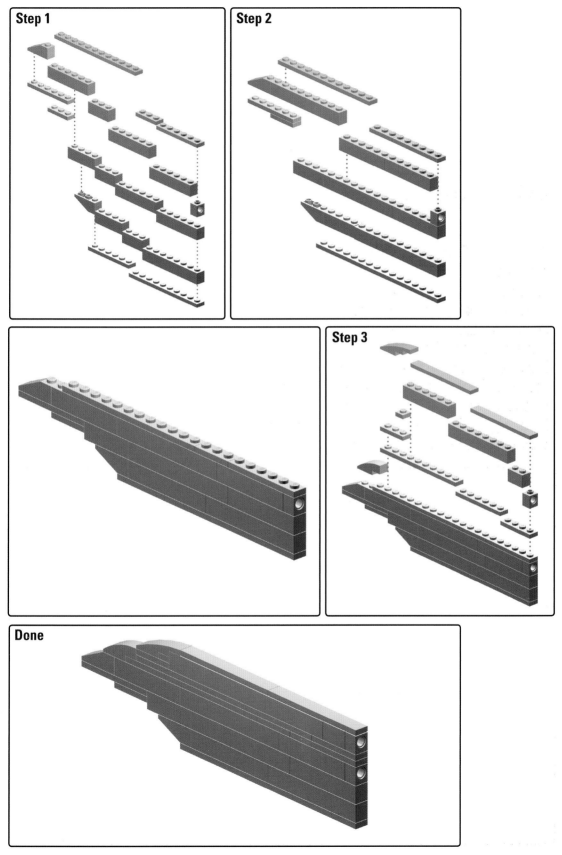

HANDLE

Step 1

Step 2

Done

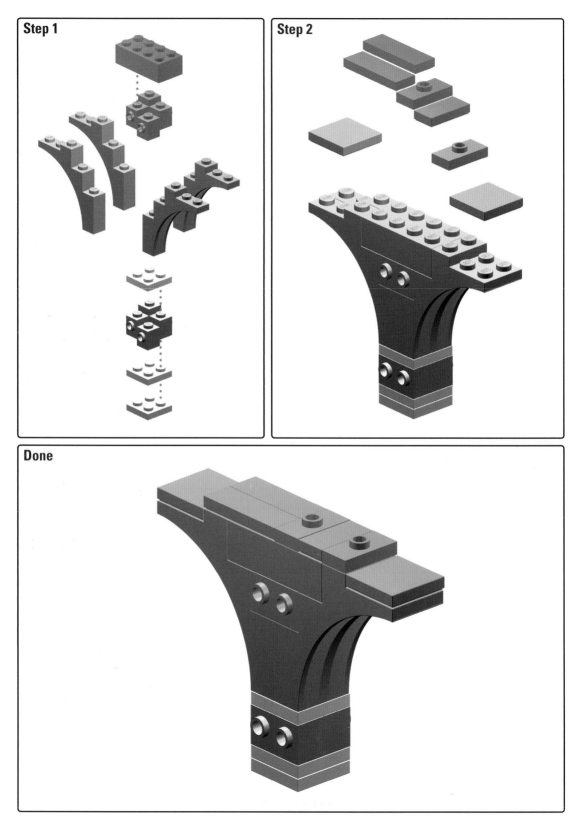

HANDLE COVERS

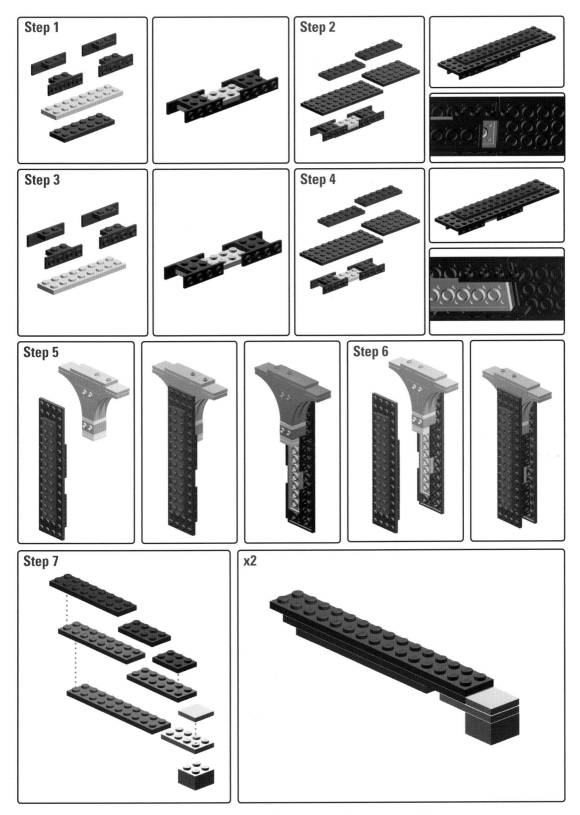

Step 1

Step 2

Step 3

Step 4

Step 5

Step 6

Step 7

x2

Step 8

Done

FINAL ATTACHMENTS

Step 1

Complete!

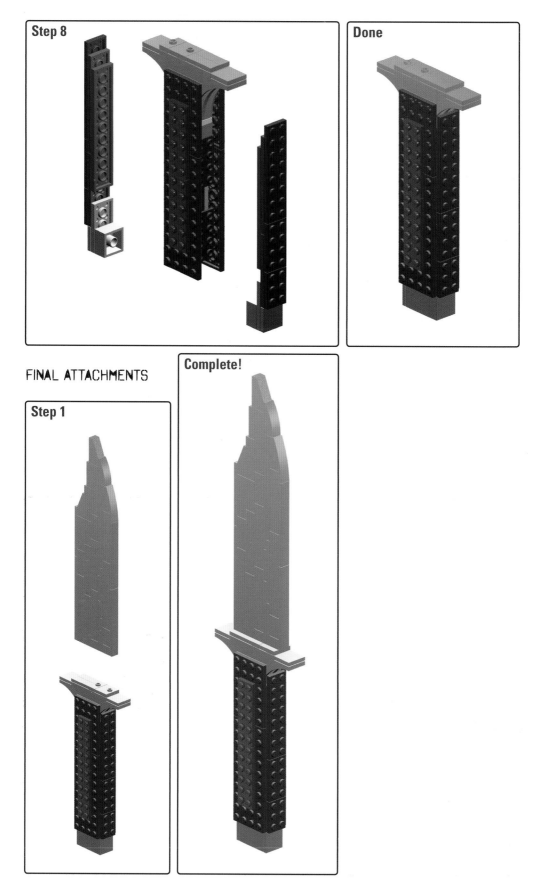

TOMAHAWK

SCALE: 1:1

This badass North American throwing weapon gets its name from the Powhatan word for cutting tool. Before European settlers brought over metal that could be crafted as a blade, Native Americans constructed their tomahawks with shafts of hickory or oak with deer antlers or rocks at the end. Powerful for throwing and cutting, this handheld axe was the everyman's weapon and tool. If wielding a deadly axe on a regular basis weren't badass enough, some tomahawk bearers carved bowls into their wooden handles so they could smoke tobacco after a long day of being the man.

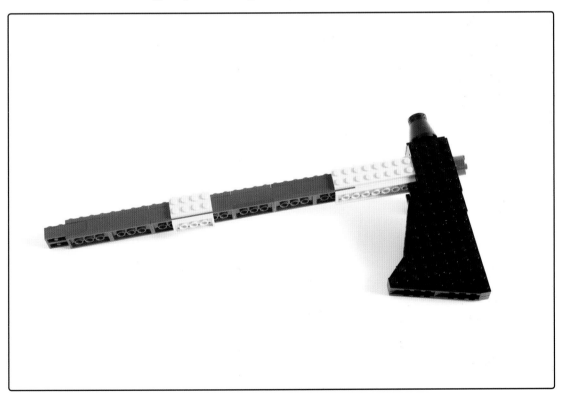

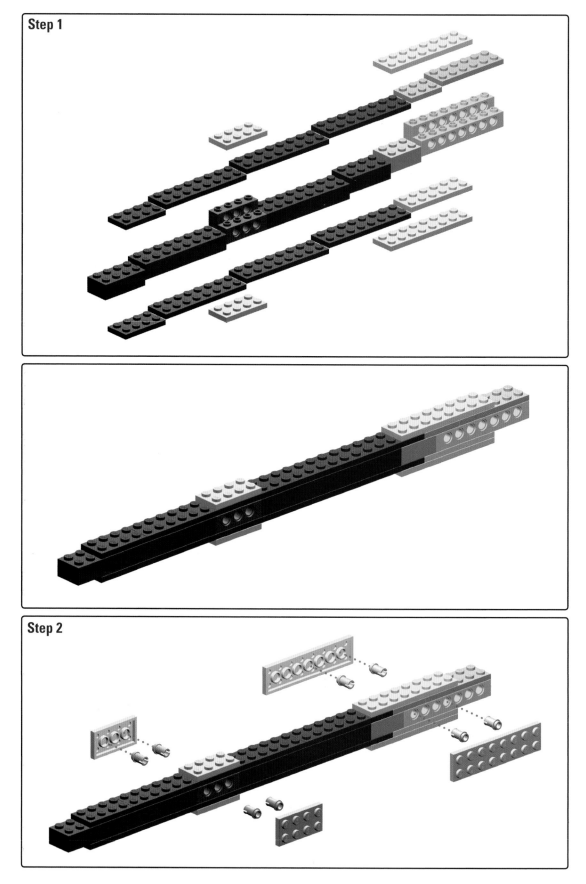

Step 1

Step 2

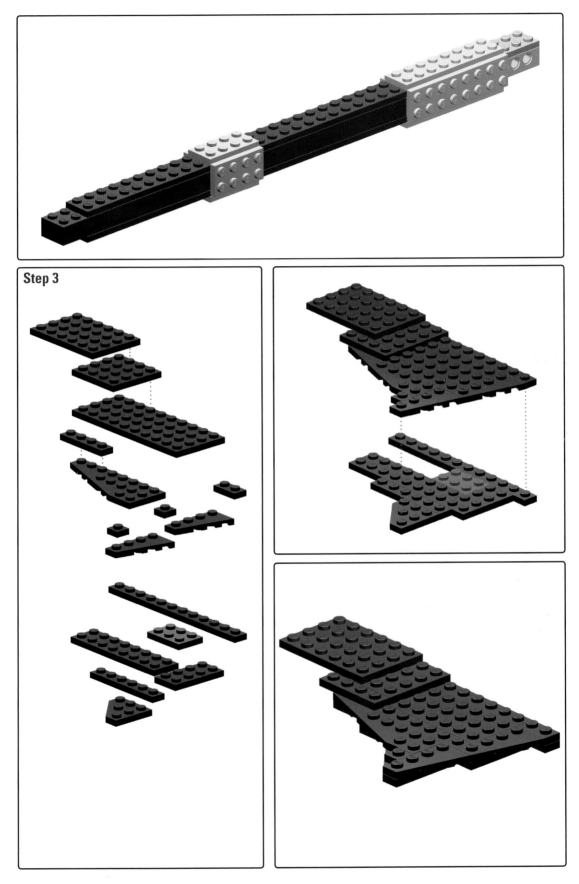

Step 3

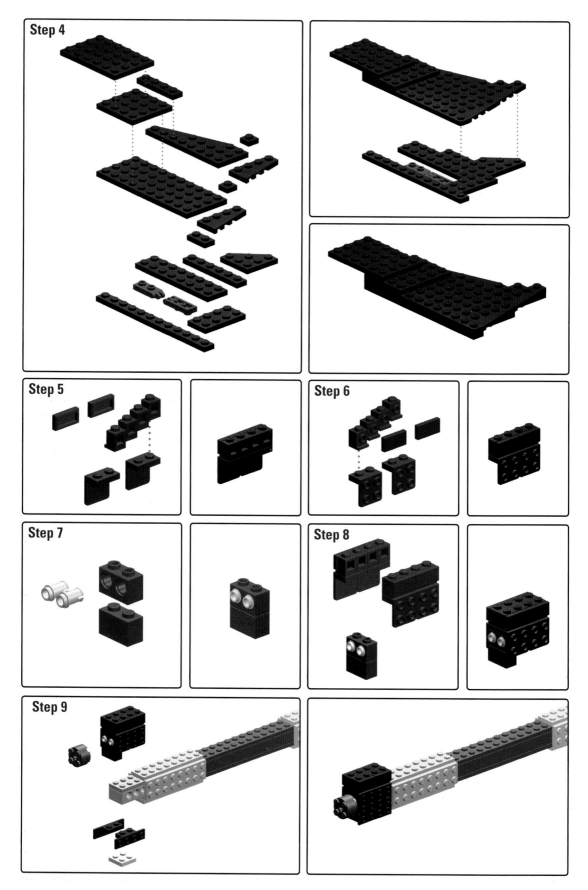

Step 4

Step 5

Step 6

Step 7

Step 8

Step 9

Step 10

Complete!

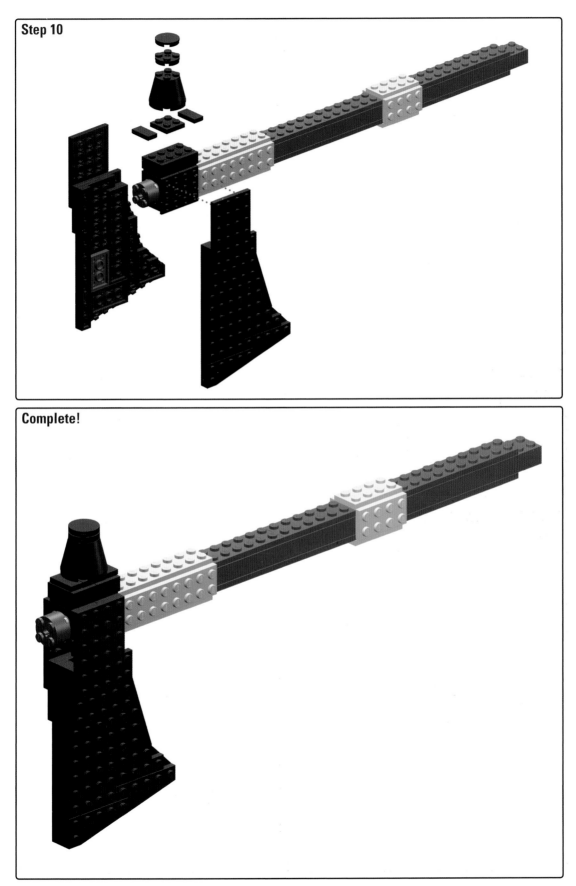

CHAPTER TWO
RANGED WEAPONS

BLOWGUN
SCALE: 1:1

This sly weapon has been used throughout history across many cultures. Made of a long wooden cylinder that shoots seeds, clay pellets, or darts, it was generally used to hunt birds and small animals. Some users of the weapon soaked the darts in poison beforehand, ensuring the kill. Blowguns can vary in size, especially depending on the origin of the gun. Fukiya, from Japan, are generally about four feet in length and do not have a mouthpiece. Cherokee blowguns, on the other hand, can range from six to nine feet long and are made of river cane. Both versions are used in global dart-blowing competitions and are currently being vetted for a place as an Olympic sport. This particular model has a smooth barrel made to fit a 1x1x4 dart.

Step 1

Step 2

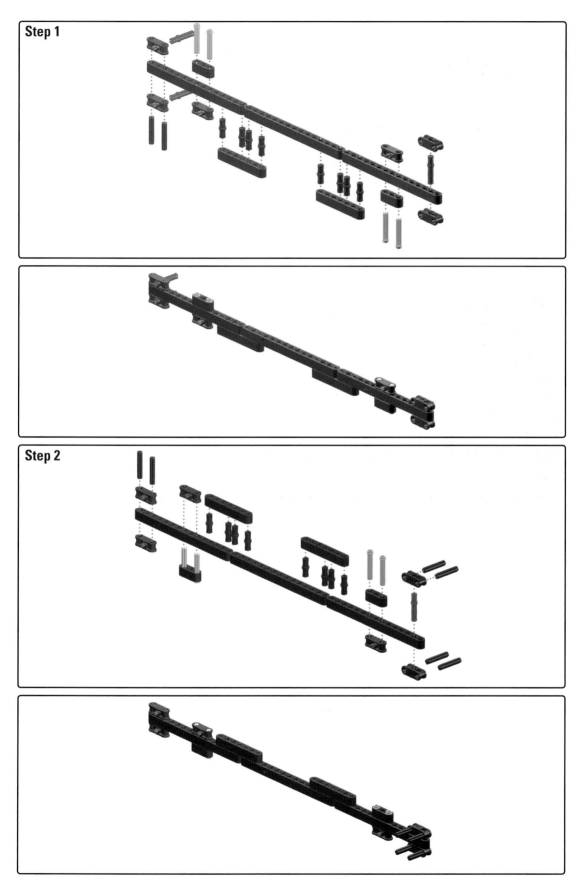

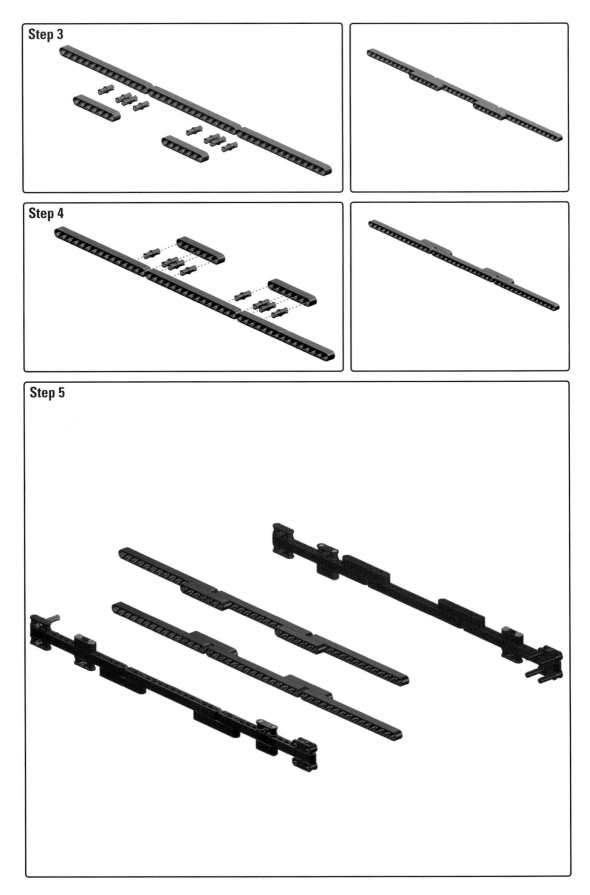

Step 3

Step 4

Step 5

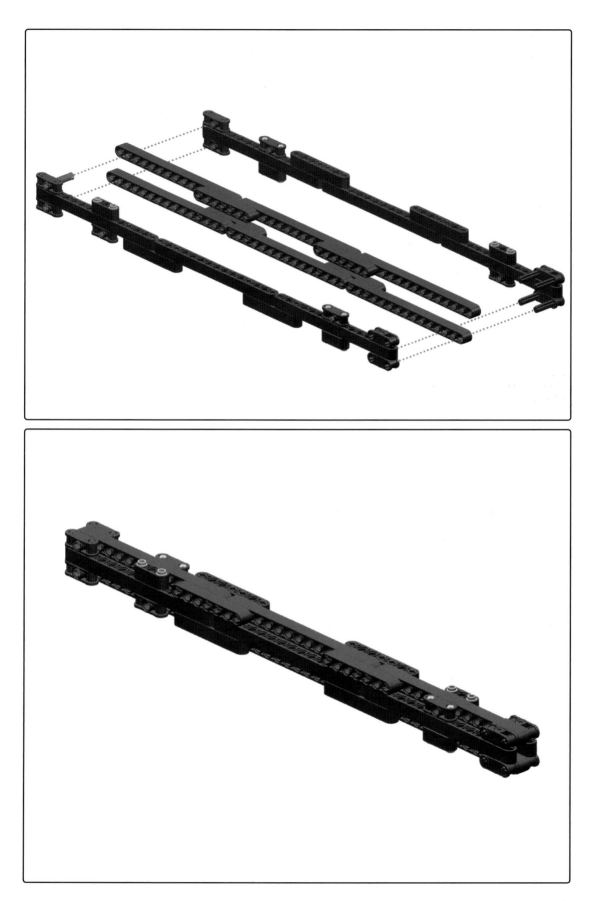

Step 6

Step 7

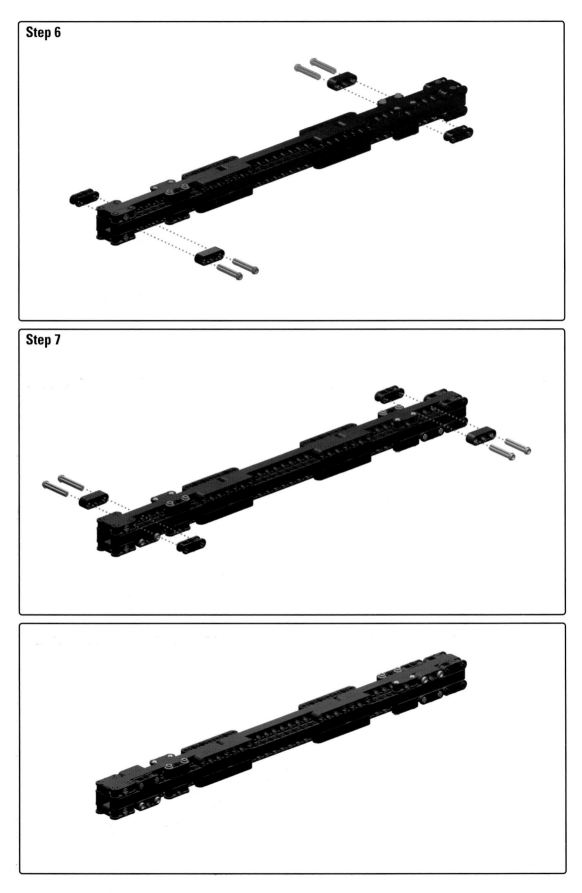

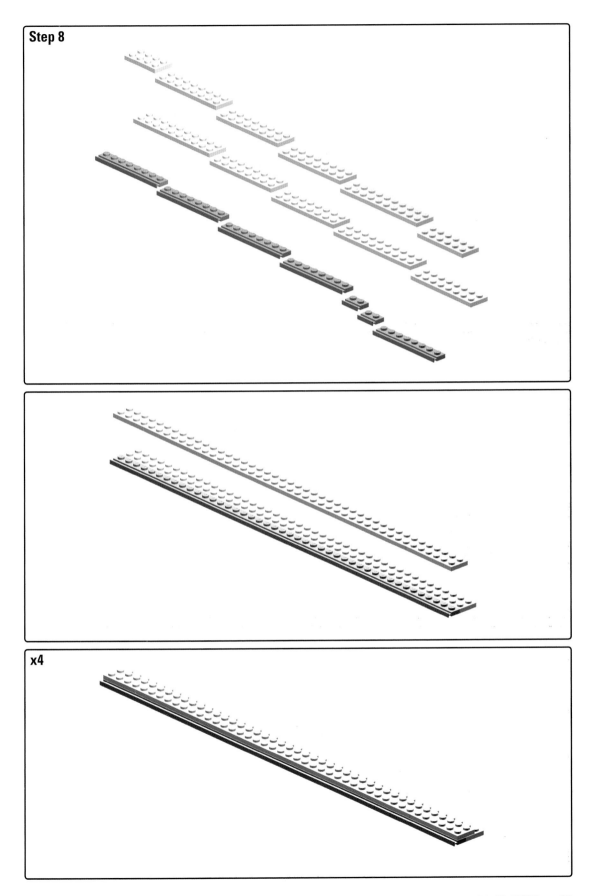

Step 8

x4

Step 9

Step 10

Complete!

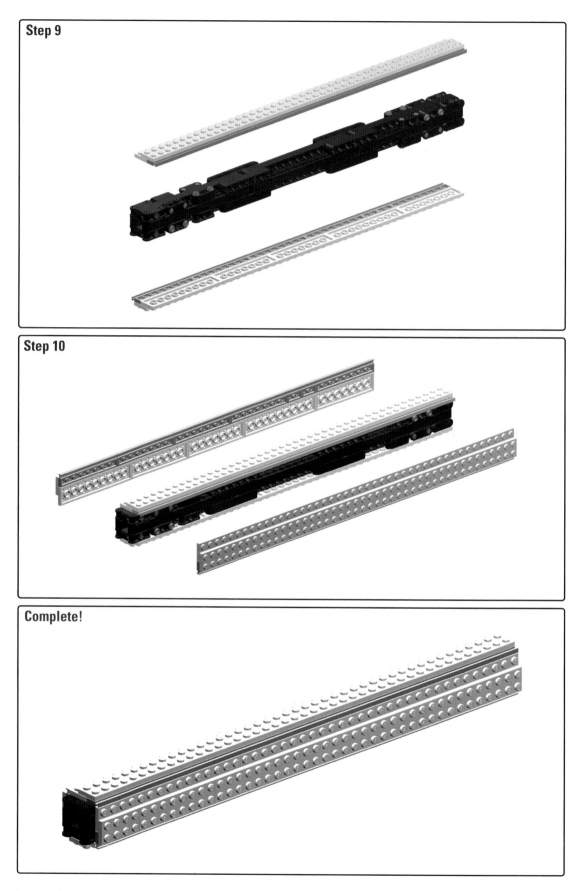

CROSSBOW PISTOL
SCALE: 1:1

The handheld crossbow has been used since late medieval times: it is the industrious man's updated bow and arrow, allowing the wielder to load the arrow or bolt and then aim, pulling a trigger to release the projectile. Modern handheld crossbows function much like a one-shot, silent pistol. This model requires one large rubber band to activate. The trigger mechanism can be wound up multiple times and is semi-automatic, meaning that if someone removes the top of the slide and the bows, it can fire as many times as you put rubber bands on.

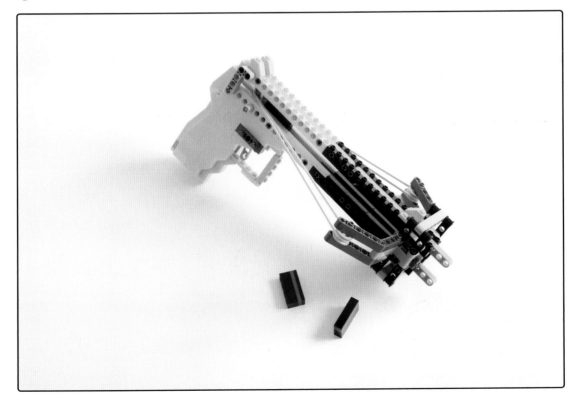

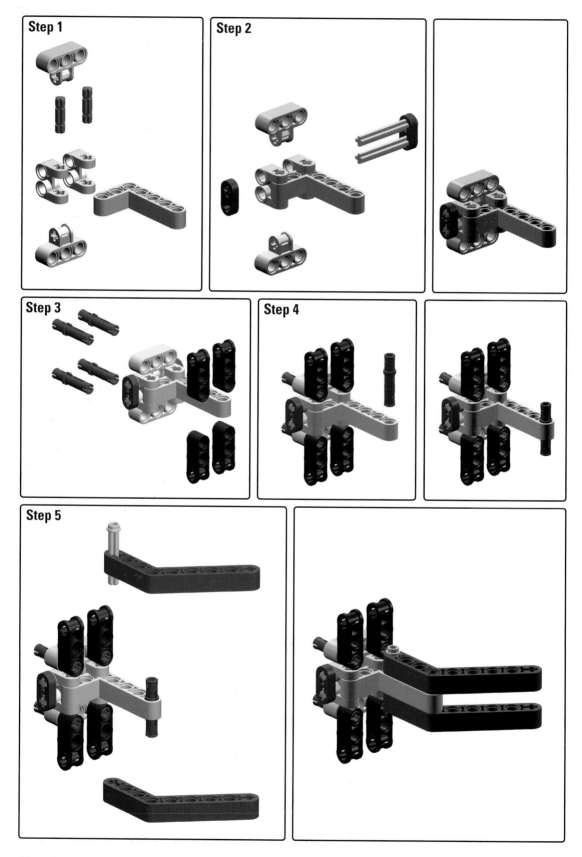

Step 1

Step 2

Step 3

Step 4

Step 5

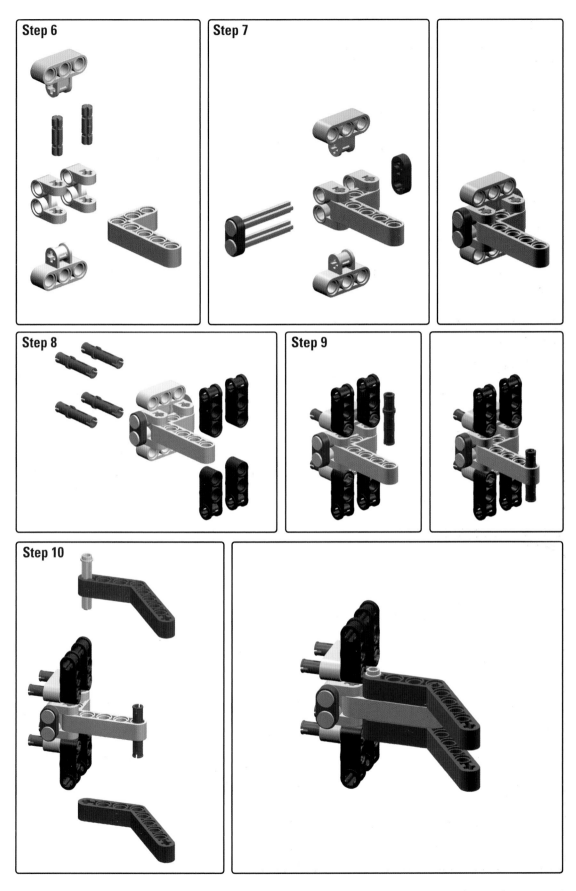

Step 6

Step 7

Step 8

Step 9

Step 10

Done

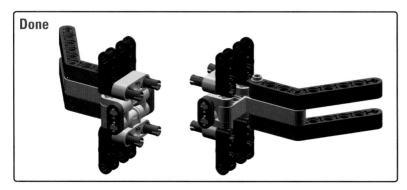

BARREL

Step 1

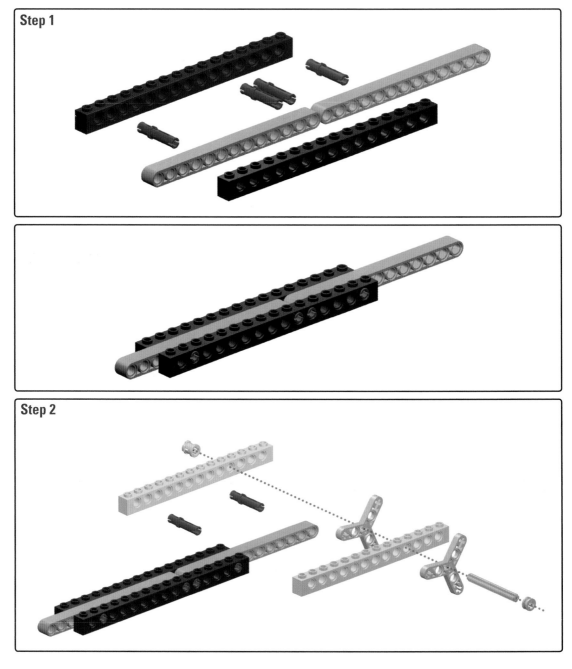

Step 2

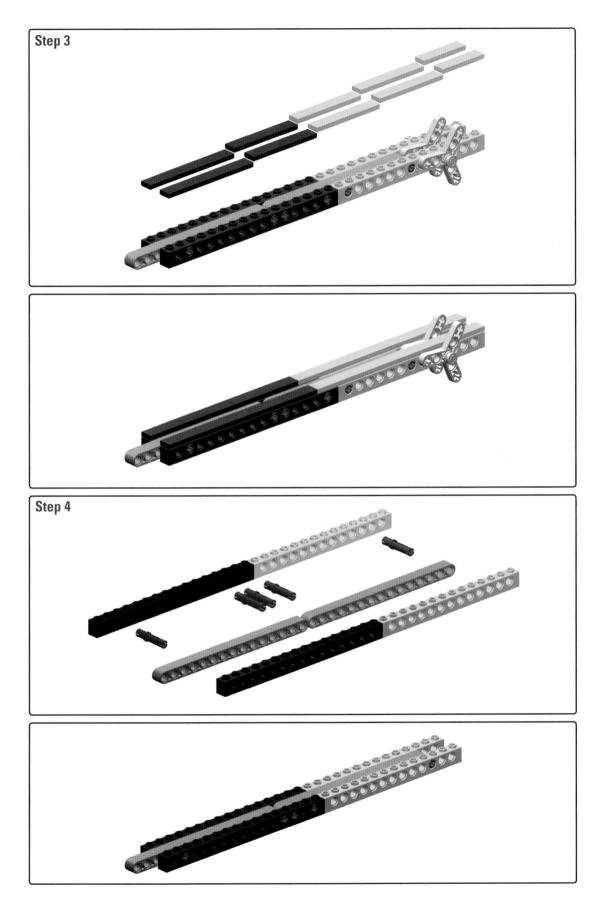

Step 3

Step 4

Step 5

Step 6

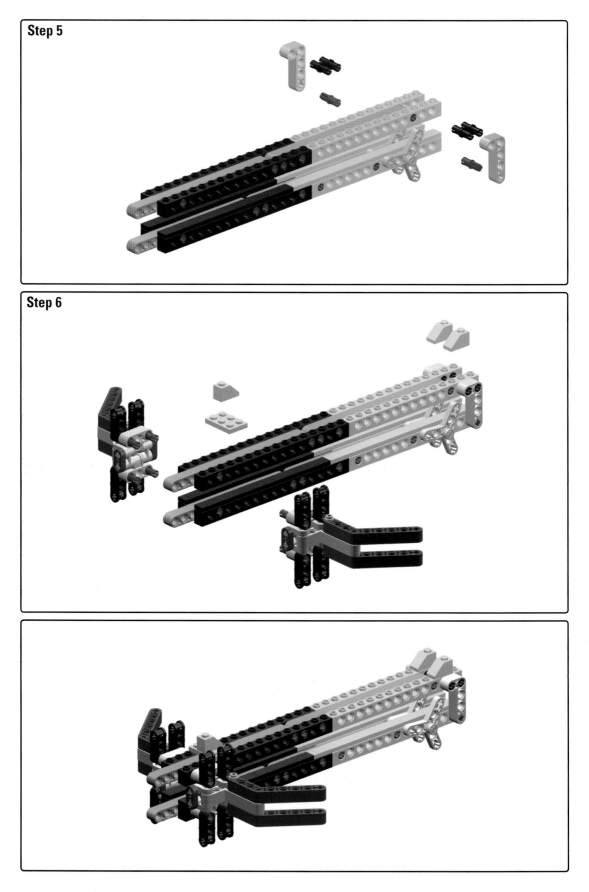

Step 7

Done

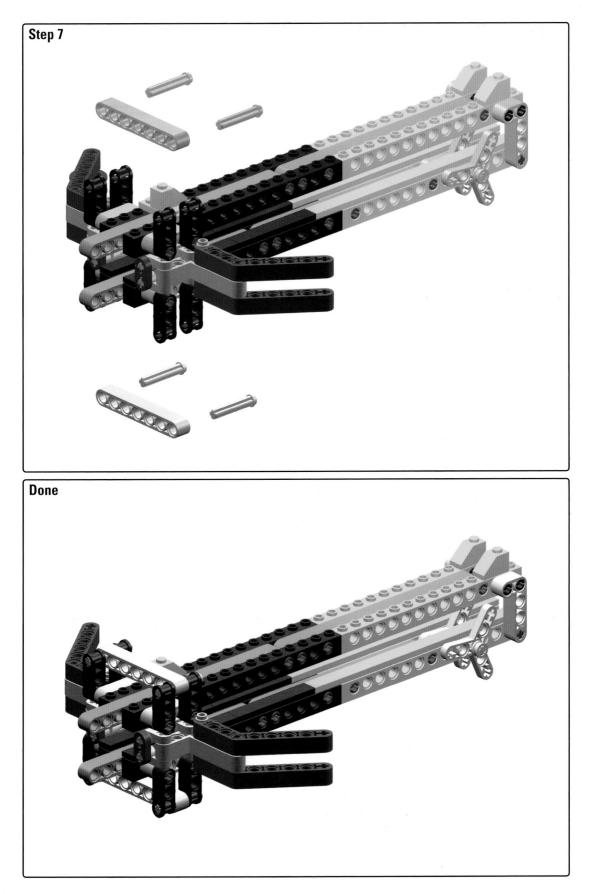

HANDLE

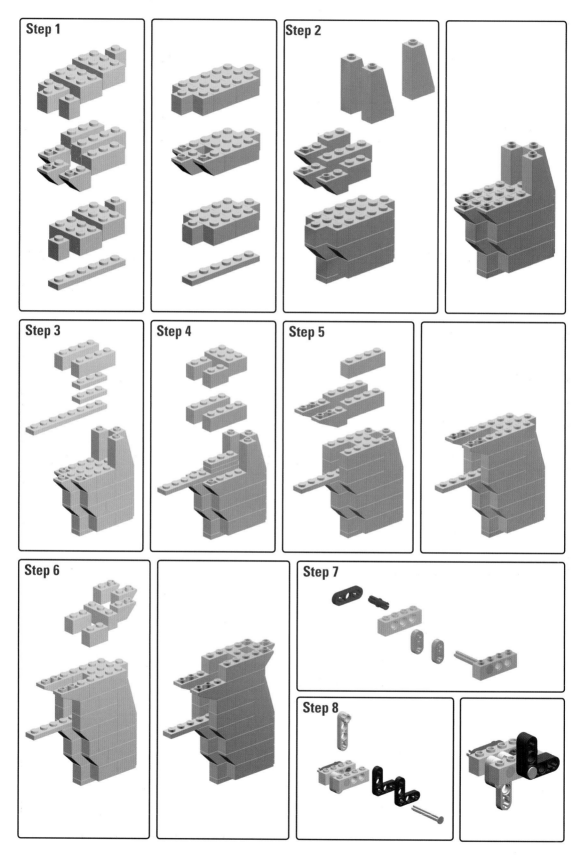

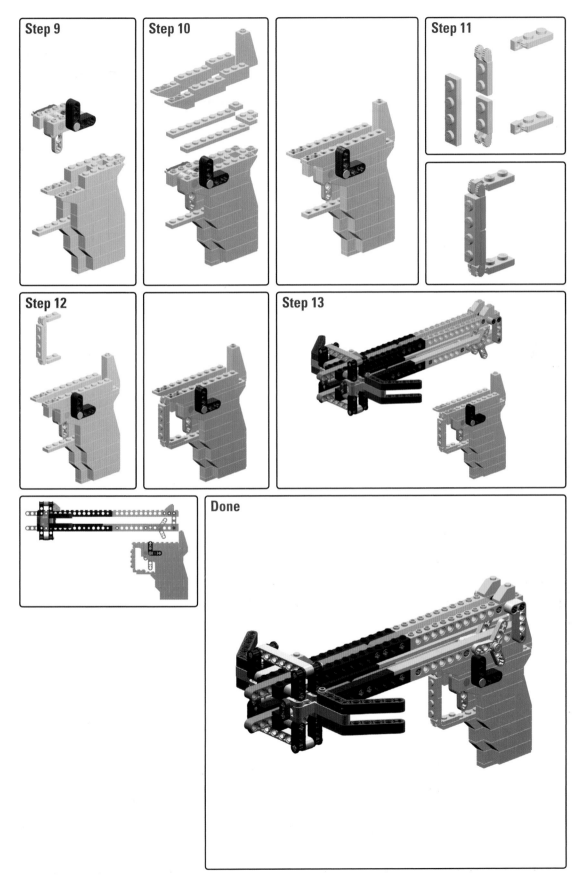

Step 9

Step 10

Step 11

Step 12

Step 13

Done

RUBBER BAND

Step 1

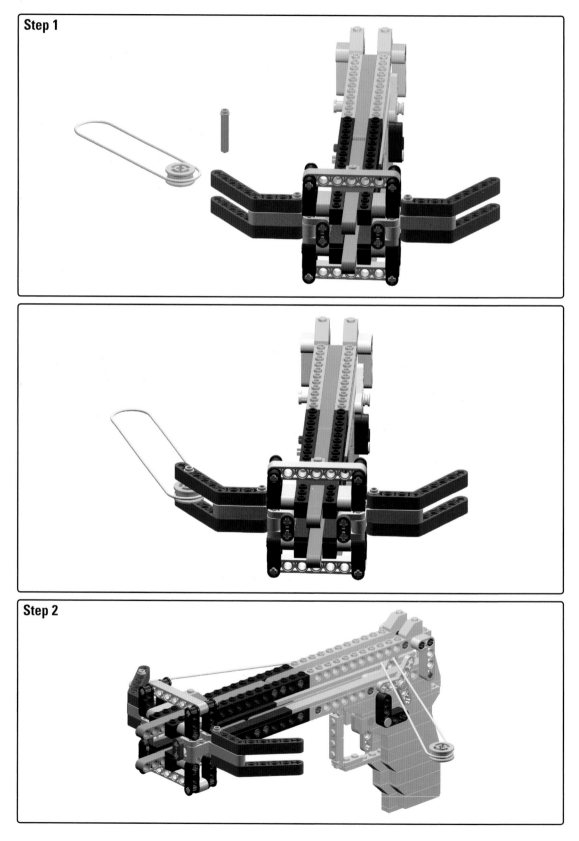

Step 2

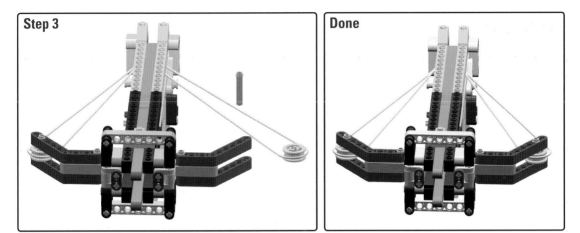

Step 3

Done

Complete!

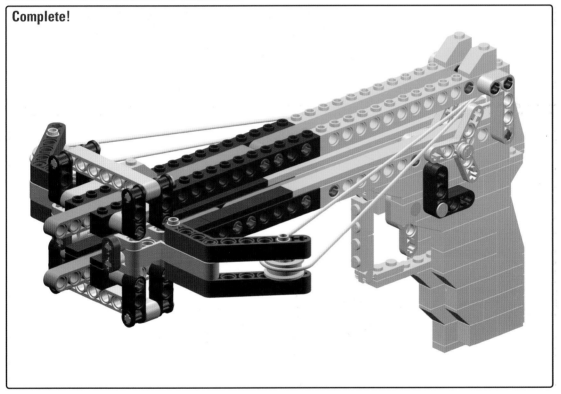

NINJA STAR

SCALE: 1:1

The term "shooting star" never sounded so deadly. A Japanese *shuriken* is a concealed, multi-edged, bladed weapon that bolstered the already intimidating arsenal of a samurai. Often used for tactical advantages of surprise or as a distraction, ninja stars come in various shapes and can be used for throwing, stabbing, or slashing. Less forceful than the *katana*, but definitely stealthier, the ninja star is the premier tool for those who are versed in the art of *shurikenjutsu*.

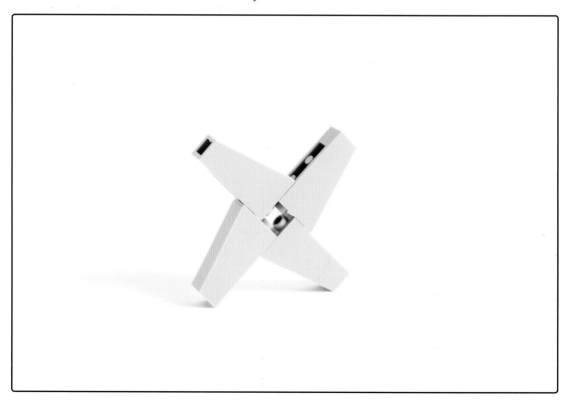

Step 1

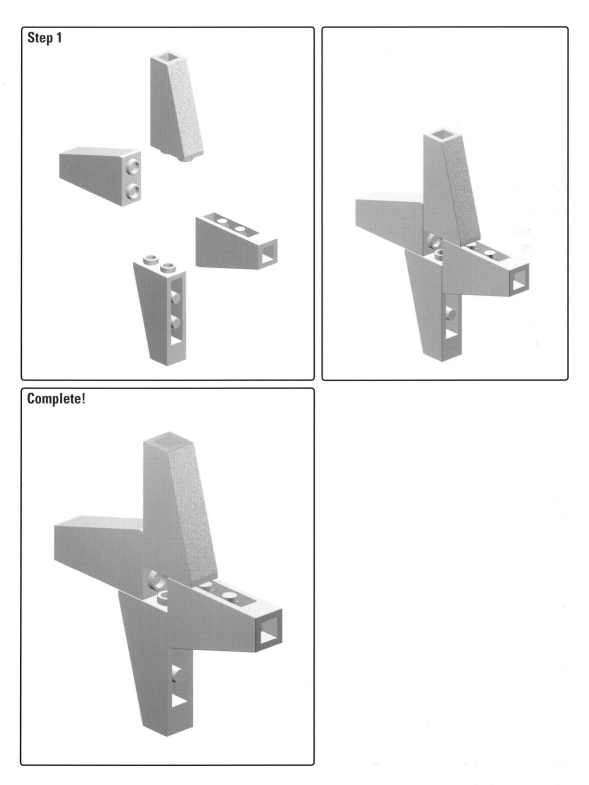

Complete!

BALLISTA

SCALE: MINI-FIGURE

This monster weapon is a crossbow on steroids. An early Greek and Roman siege weapon, a ballista was a large, mostly wooden structure that used torsion from springs to launch a projectile to a doomed target—in other words, it was the ancient equivalent to a guided missile. The framework for the ballista was later used to create a smaller sniper weapon called the Scorpio, which was a terrifyingly accurate catapult system that led to the invention of the crossbow.

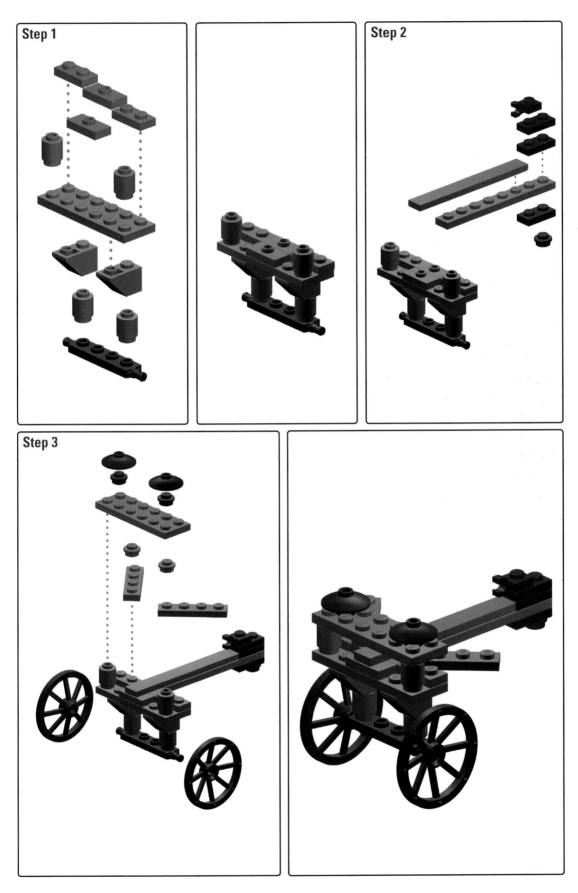

Step 1

Step 2

Step 3

Step 4

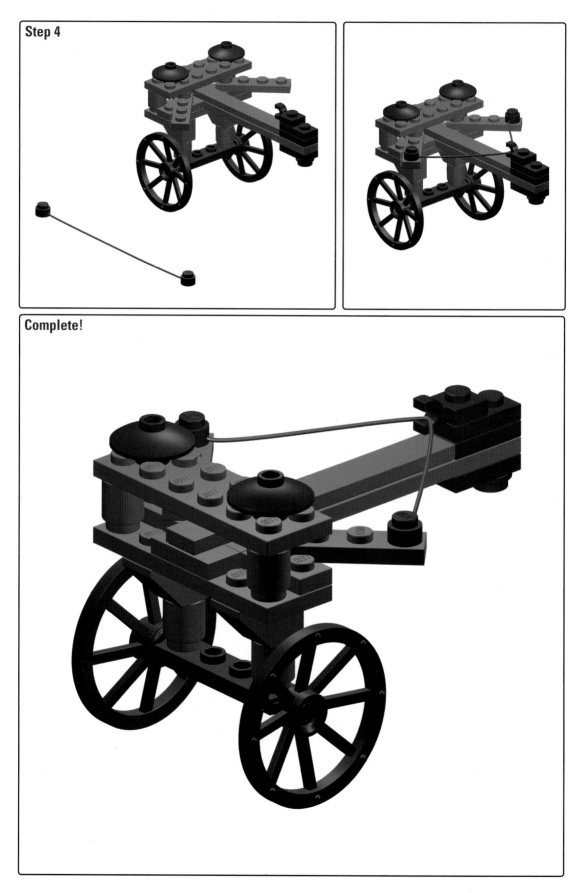

Complete!

CANNON

SCALE: MINI-FIGURE

This American Civil War–era cannon
was a titan of force in its day. Bigger,
more destructive, and more accurate
than its predecessors, this cannon
could launch projectiles over a mile
away. As a method of deception, some
soldiers used painted logs to create
"Quaker Guns," which mimicked
the look of a real cannon to confuse
enemy troops. Occasionally, these
enemies were unlucky enough to
have real cannons put into the mix.

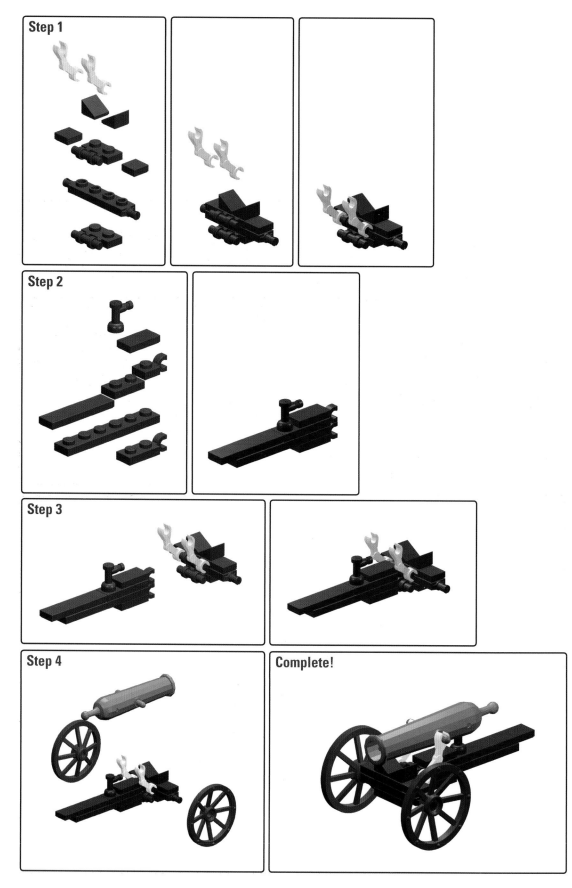

Step 1

Step 2

Step 3

Step 4

Complete!

TOMMY GUN
SCALE: MINI-FIGURE

It's no wonder this submachine gun was originally called "The Annihilator." A compact, reliable weapon that used .45 ACP cartridges and a box magazine, the 1919 John T. Thompson sprayed about 1,200 rounds per minute, an impressively high rate for the time. The infamous Tommy gun was originally fashioned as a military-grade "trench sweeper" but became the favored weapon of law enforcement, gangsters, and bootleggers alike. Because of the ubiquitous media presence, the Tommy gun became a symbol of Prohibition-era subversive culture.

Step 1

Complete!

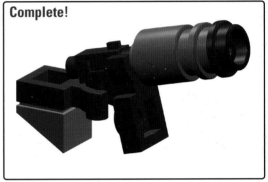

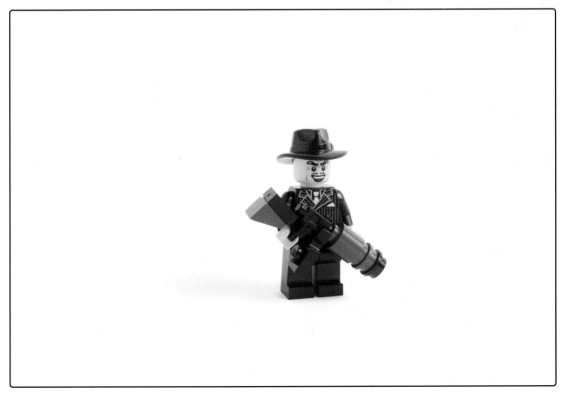

GATLING GUN

SCALE: MINI-FIGURE

The mob boss of early rapid-fire weapons, the Gatling gun was an 1862 creation of Dr. Richard Gatling for the American Civil War. While the system itself looks like a cannon with an identity crisis, courtesy of its double wheelbase, the firing system was a hand-cranked, multi-barrel system. Its inventor intended the rapid-fire military weapon to reduce the number of soldiers in war and put a spotlight on the futility of bloodshed. Ironically, it was so baller that it became the progenitor of the modern machine gun.

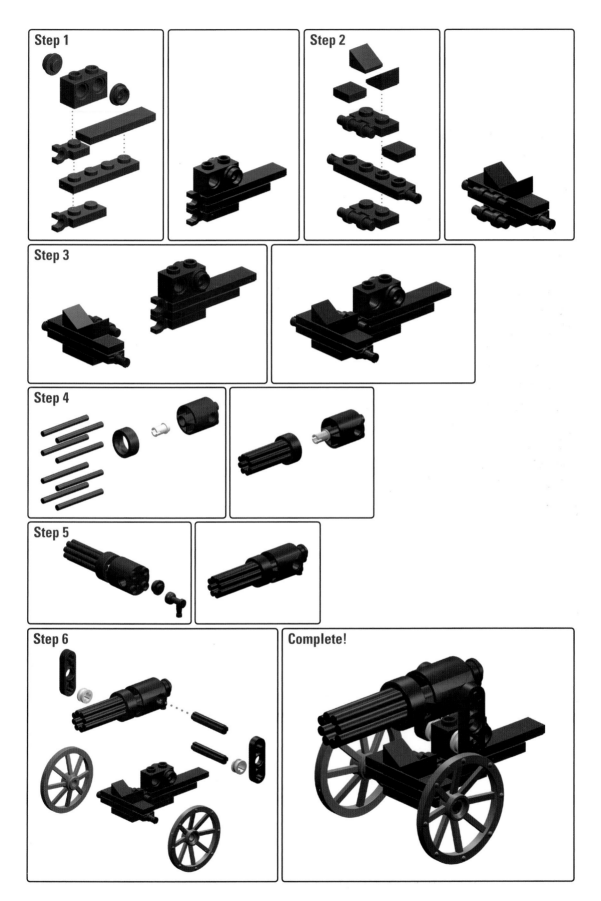

WALTHER PPK
SCALE: 1:1

In service since 1931, this sleek gun is a .32 ACP caliber blowback-operated semi-automatic pistol. Walther being a German company, the title PPK stands for *Polizeipistole Kriminalmodell*, or Police Pistol Detective Model. With a smooth, easily concealable design, this gun is the optimal piece of weaponry for undercover work. Unsurprisingly, it is best known as James Bond's weapon of choice.

HANDLE

Step 1

Step 2

Step 3

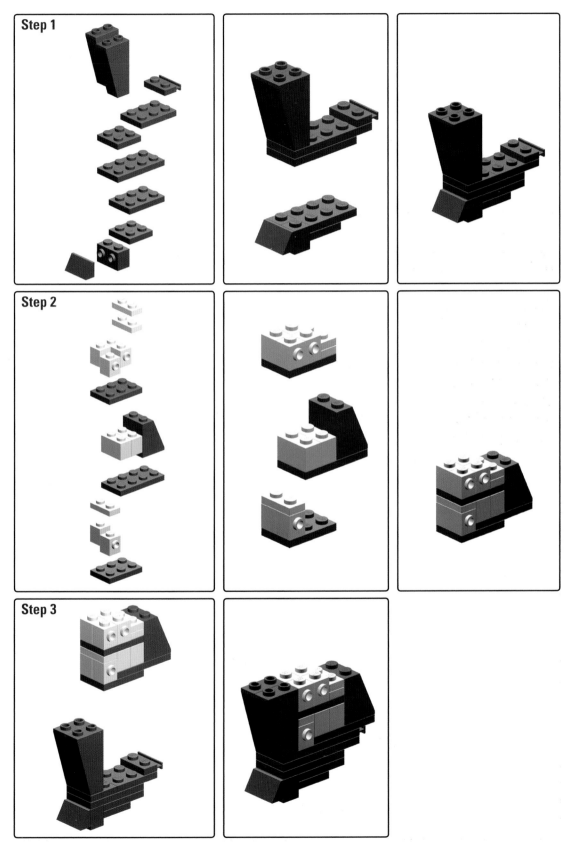

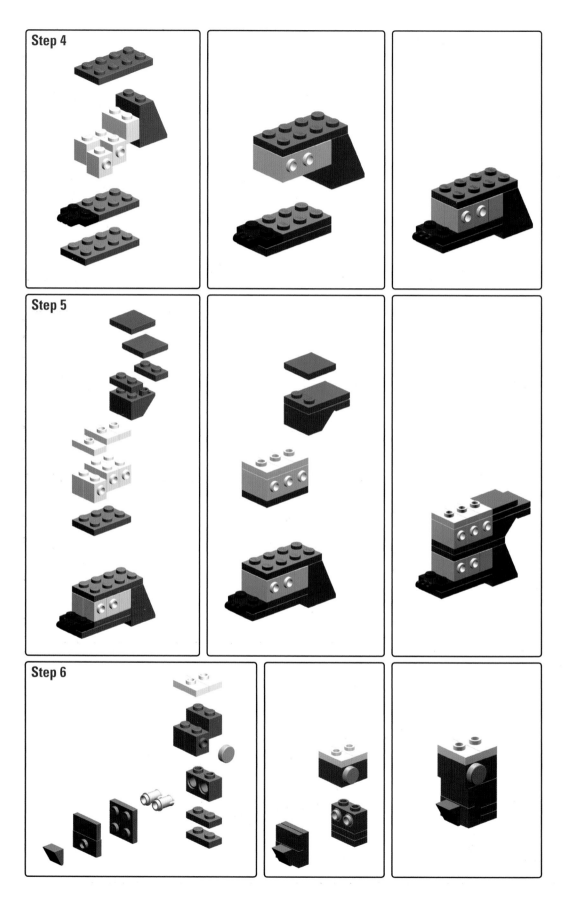

Step 4

Step 5

Step 6

Step 7

Step 8

Done

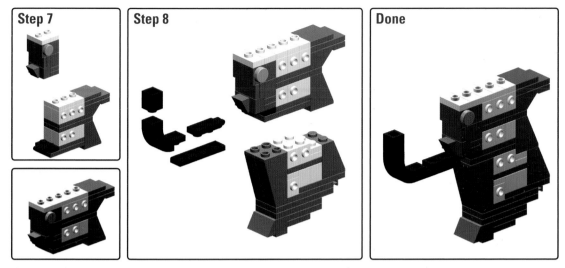

BARREL

Step 1

Step 2

Step 3

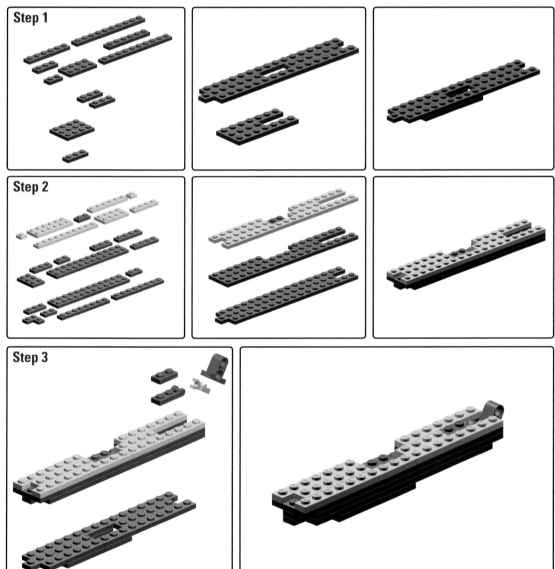

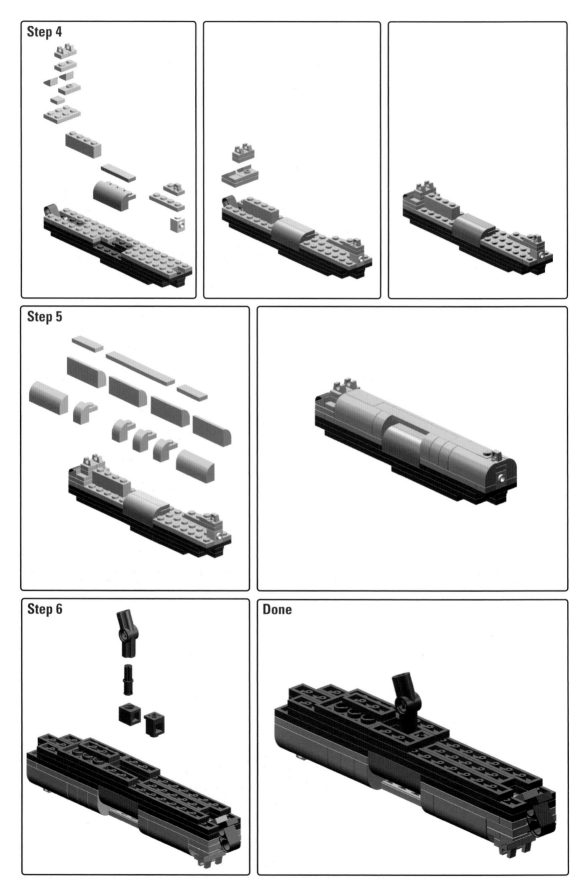

Step 4

Step 5

Step 6

Done

FINAL ATTACHMENTS

Step 1

Step 2

Step 3

Complete!

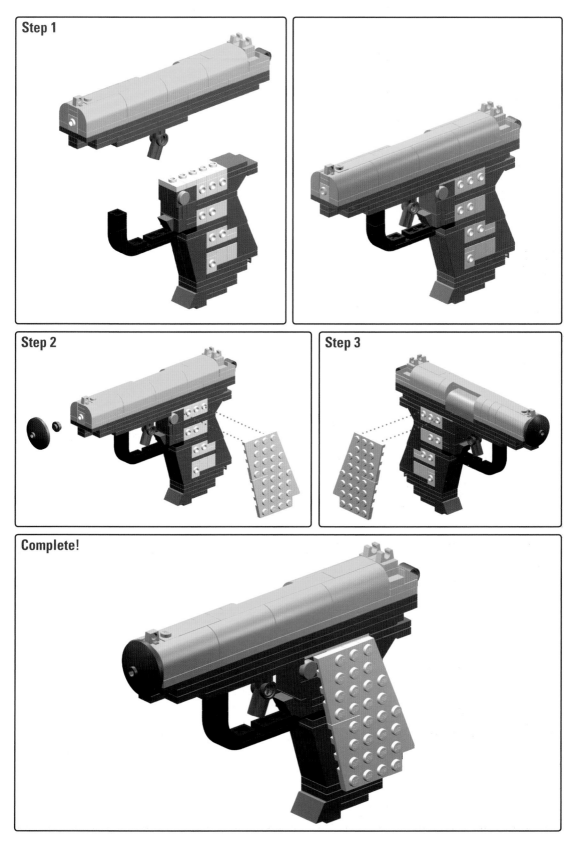

CHAPTER THREE
HEAVY ASSAULT

SIEGE TOWER

SCALE: MINI-FIGURE

Because overthrowing your neighbor could be so easy. A siege tower was widely used in medieval battles when approaching the defensive walls of an enemy. Generally as tall or taller than the enemy walls, the siege tower helped the crossbow-, pike-, and sword-bearing soldiers attack with greater advantage and cross over the walls in full force. Because they were giant wooden structures on wheels, they were typically constructed on site, and they had sufficient weaponry inside to defend against catapult attacks and the like.

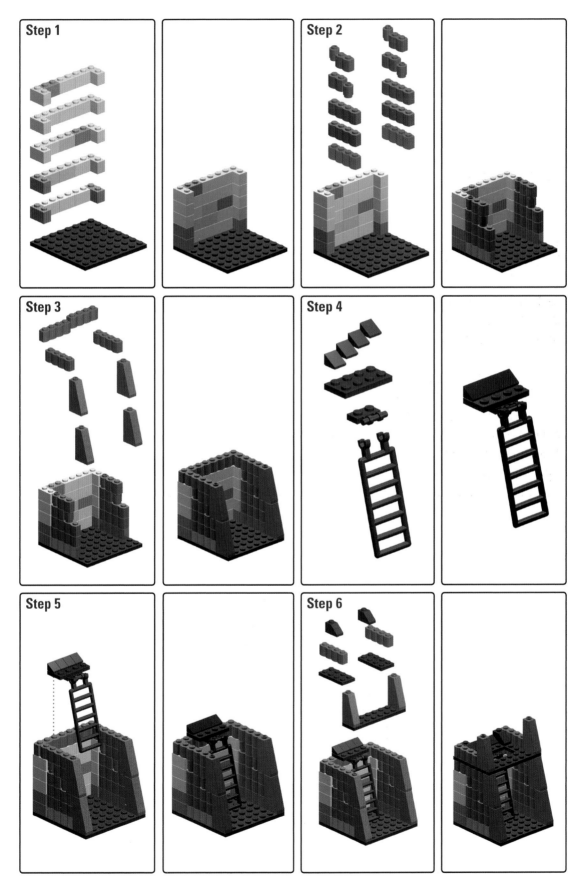

Step 1

Step 2

Step 3

Step 4

Step 5

Step 6

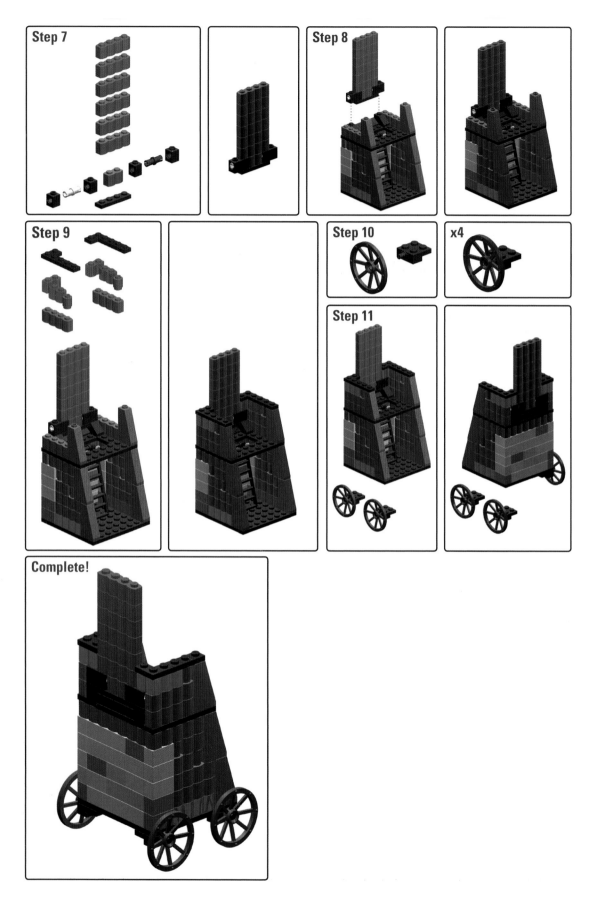

Step 7

Step 8

Step 9

Step 10

x4

Step 11

Complete!

TREBUCHET

SCALE: MINI-FIGURE

French for "greatest slingshot of all time," this weapon of mass destruction dates back to the Middle Ages. As a siege engine, it acted as a catapult for launching objects as big as 350 pounds into enemy forts or buildings. By use of a counterweight, the long arm of the trebuchet holds a projectile that flings into the air when the counterweight is triggered. Trebuchets mimicked siege towers in their wooden appearance and wheelbase and were probably the last thing anyone wanted to see creeping over the horizon.

BASE

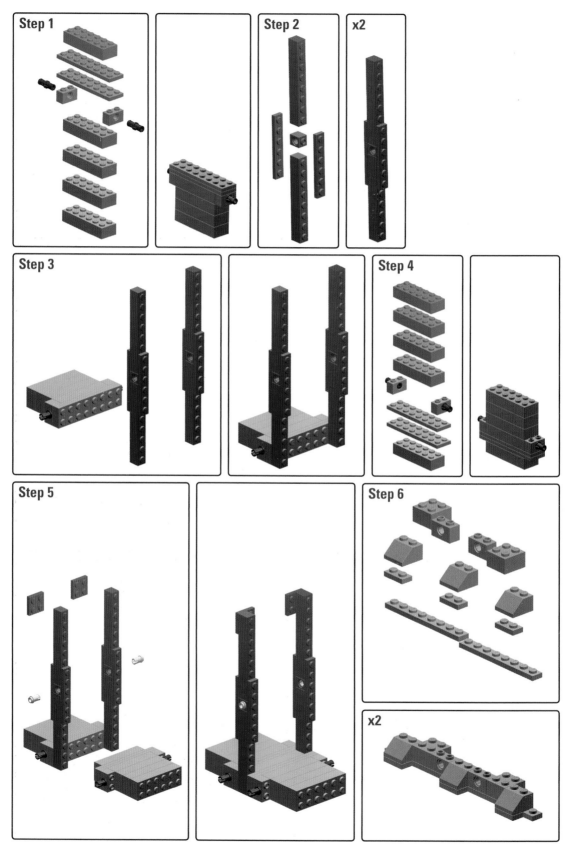

Step 1

Step 2

x2

Step 3

Step 4

Step 5

Step 6

x2

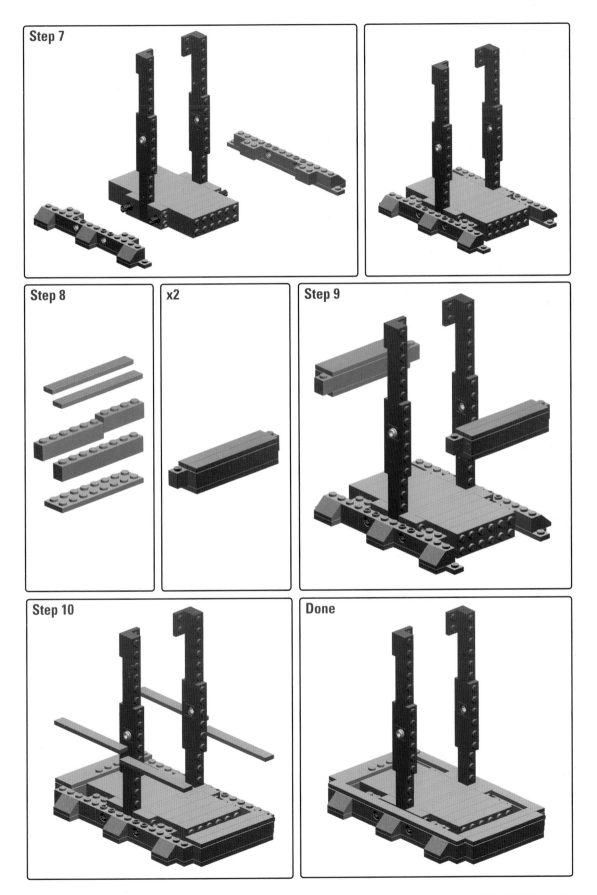

Step 7

Step 8

x2

Step 9

Step 10

Done

ARM

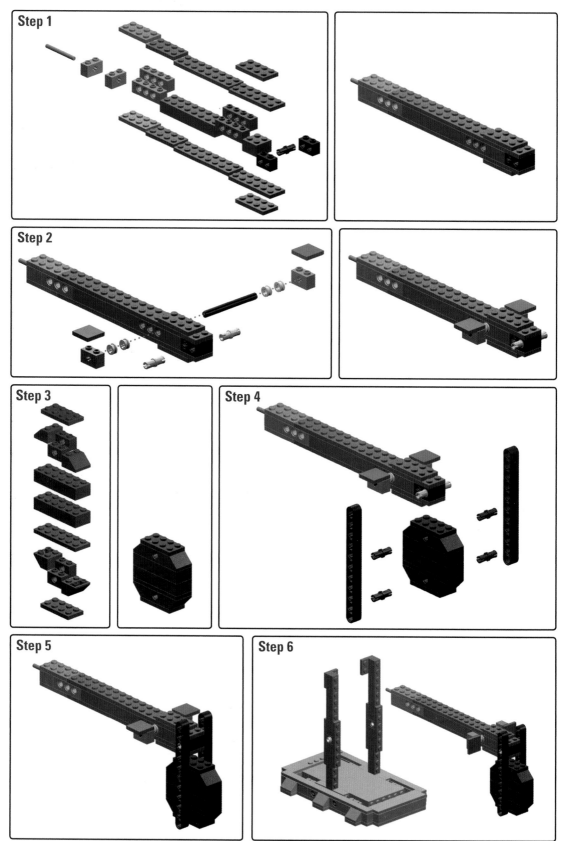

Step 1

Step 2

Step 3

Step 4

Step 5

Step 6

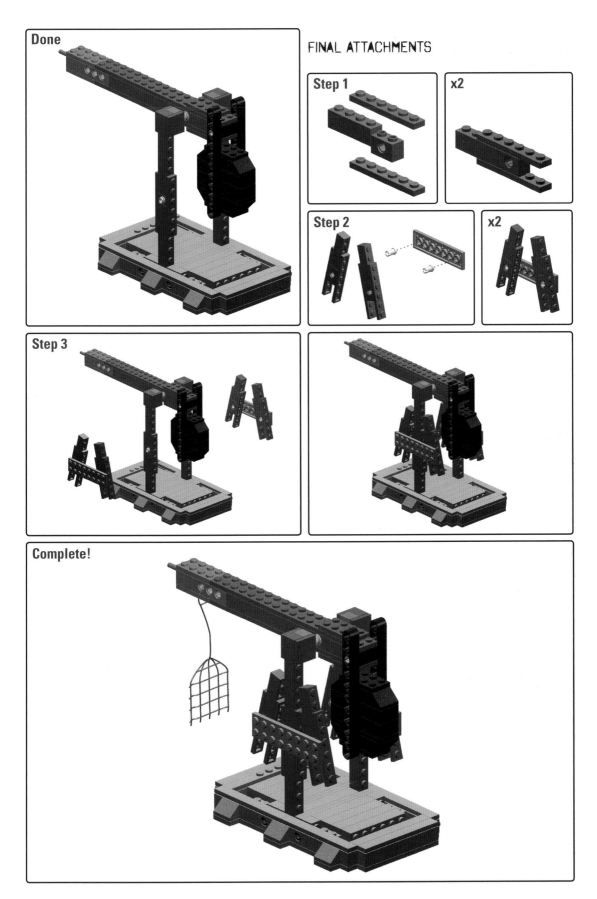

Done

FINAL ATTACHMENTS

Step 1

x2

Step 2

x2

Step 3

Complete!

BATTERING RAM
SCALE: MINI-FIGURE

When you can't find a key, use a battering ram. In its simplest form, this medieval siege weapon was a large log that could be propelled through walls, doors, or buildings by an onslaught of hulking soldiers. More complex forms have the massive log encased in a large canopy on a rolling structure that was resistant to fire and arrows, which meant bad news for anyone in its path. This dominator of all things brick and stone was the go-to weapon of force up until gunpowder made its way onto the battlefield.

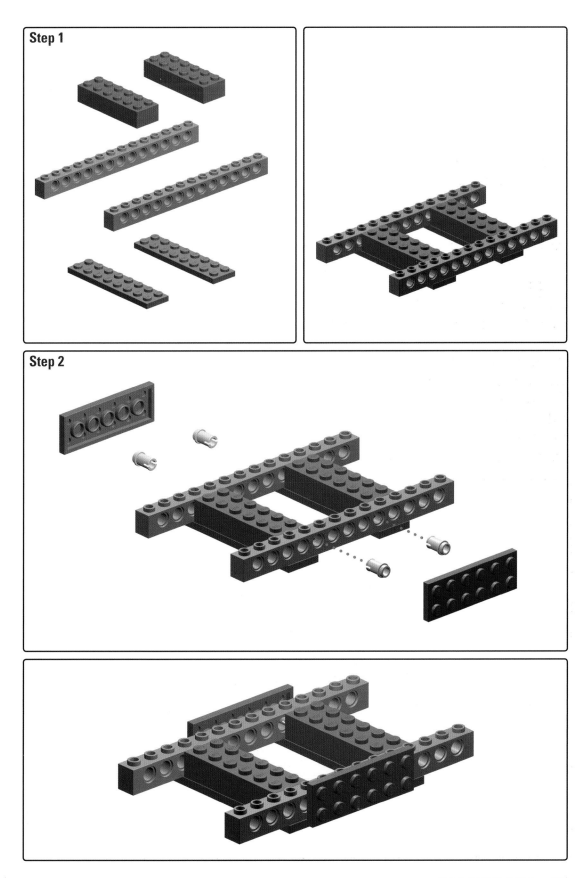

Step 1

Step 2

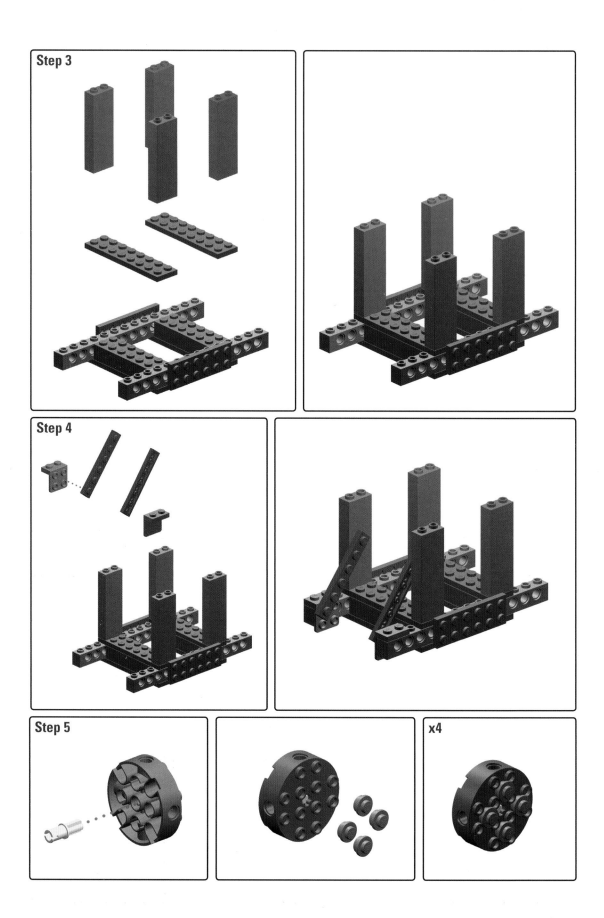

Step 3

Step 4

Step 5

x4

Step 6

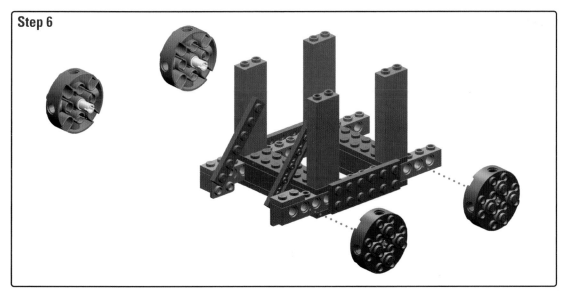

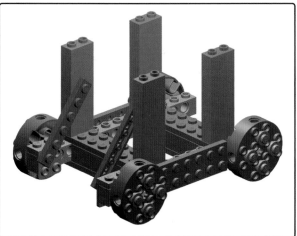

Step 7

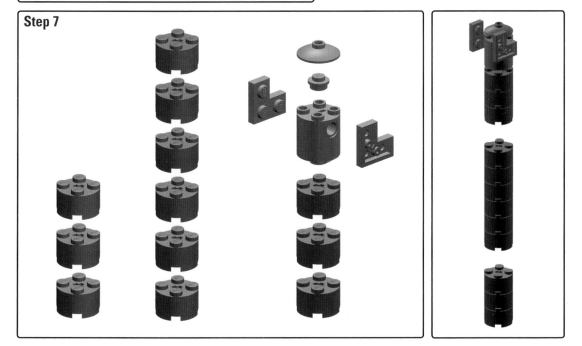

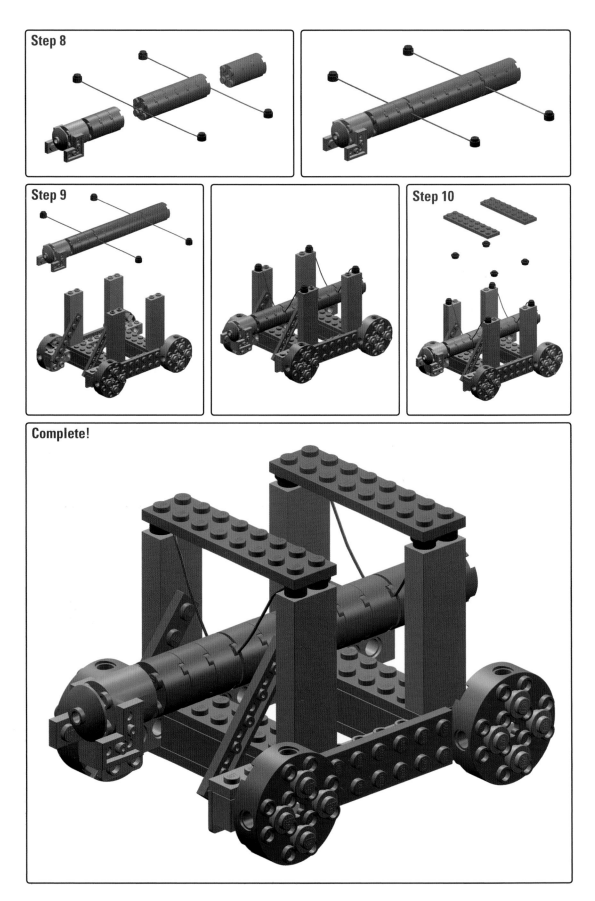

Step 8

Step 9

Step 10

Complete!

M198 HOWITZER
SCALE: MINI-FIGURE

If Rambo designed a catapult, this is what you'd get. This beast of an artillery piece is medium in size but needs about eleven soldiers to man it. This model went into full service in the US army and Marine Corps in 1979, but the concept goes back to the seventeenth century. The M198 Howitzer fires a maximum of four rounds of 155mm shells per minute through a gun tube that is stabilized over a baseplate and can launch projectiles from a range of 20,000 to 33,000 yards with guided ammunition.

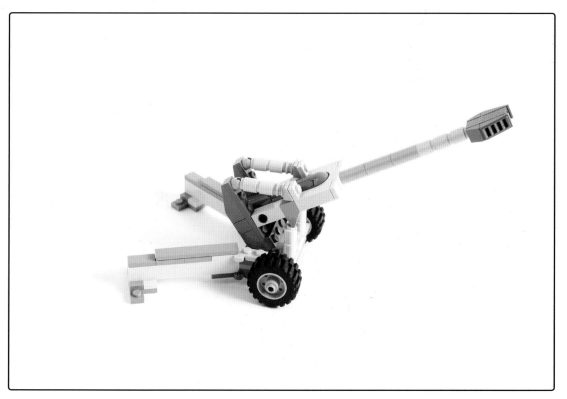

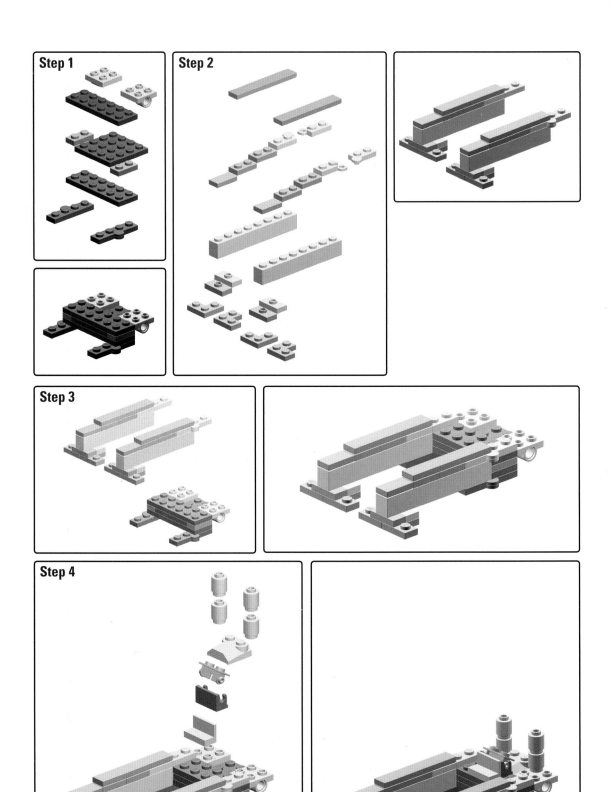

Step 1

Step 2

Step 3

Step 4

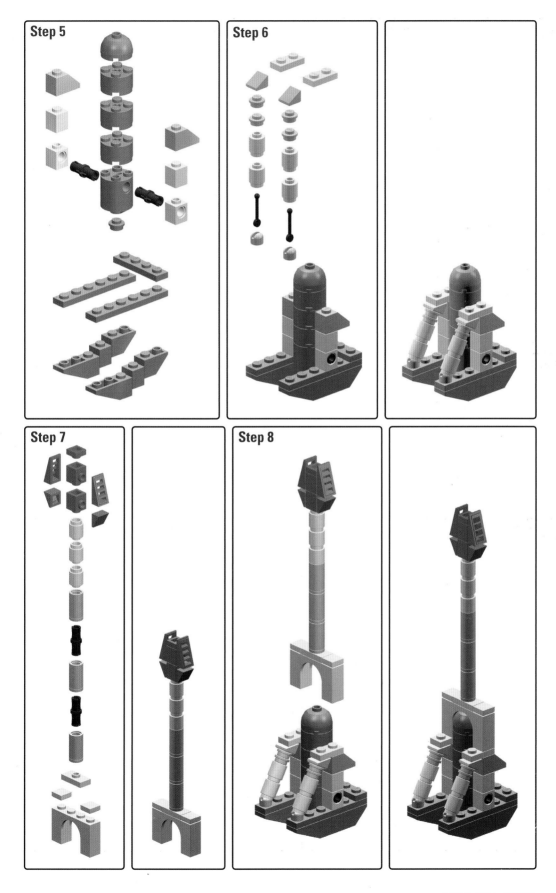

Step 5

Step 6

Step 7

Step 8

Step 9

Step 10

Complete!

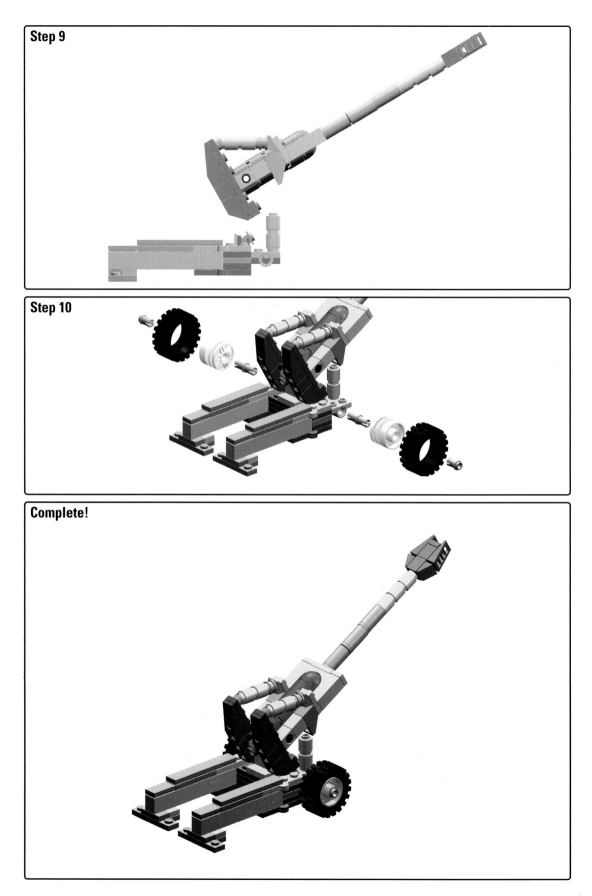

M1114 HUMVEE
SCALE: MINI-FIGURE

This SUV is so badass, Arnold Schwarzenegger
personally convinced its manufacturers
to build the Hummer for civilian sale. The
M1114 High Mobility Multipurpose Wheeled
Vehicle (HMMWV) is an all-terrain military
vehicle tricked out with heavy artillery. The
original Humvees were first serviced in the
early 1980s, but the M1114 came out in
1996 and was "up-armored" in 2004 with a
Boomerang anti-sniper detection system,
weapons mount for shooting heavy machine
guns, grenade launchers and other armament,
and CROWS, which allows the gunner to shoot
from inside the truck.

FRONT

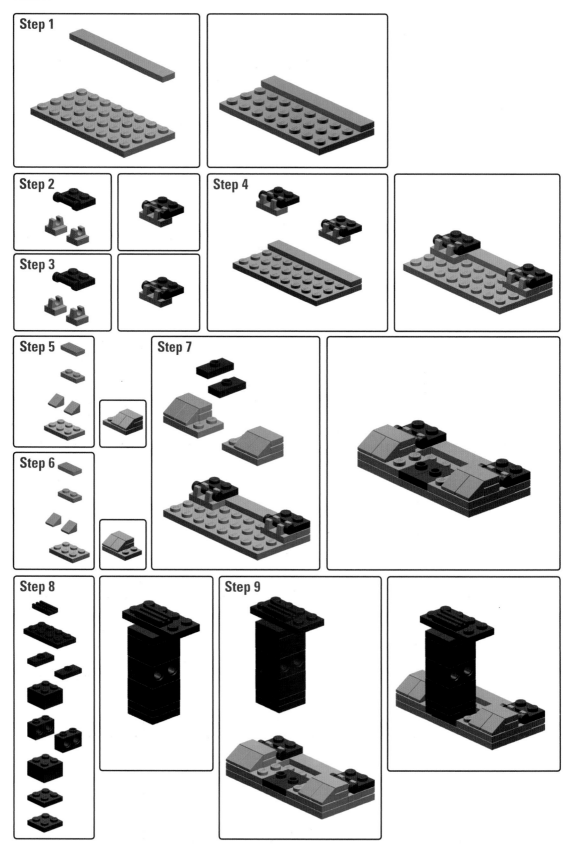

Step 1

Step 2

Step 3

Step 4

Step 5

Step 6

Step 7

Step 8

Step 9

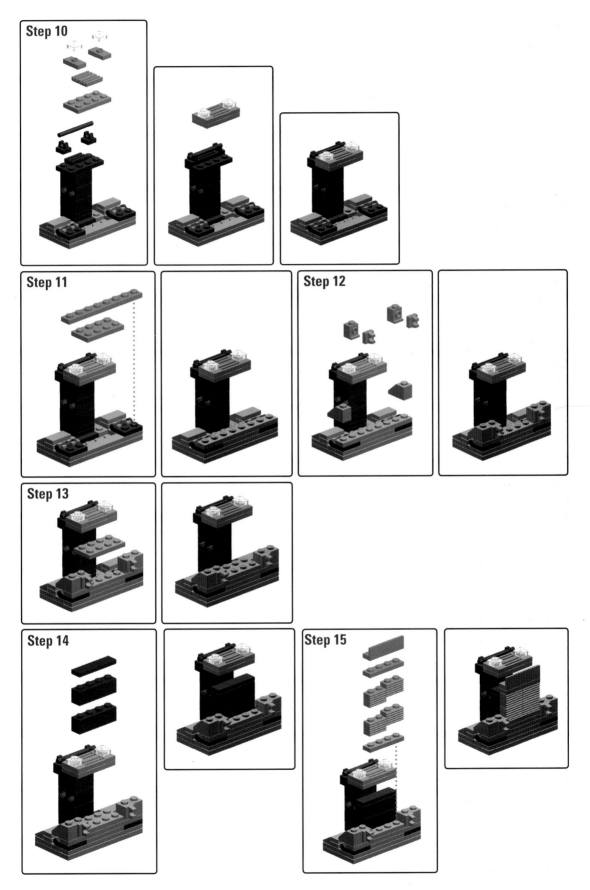

Step 10

Step 11

Step 12

Step 13

Step 14

Step 15

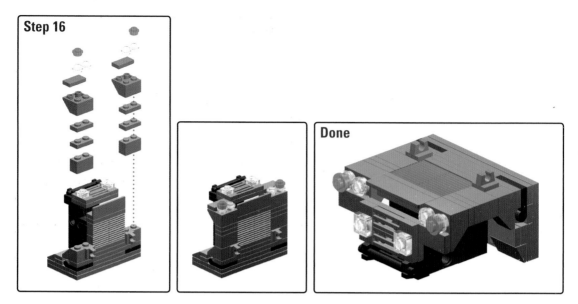

BACK

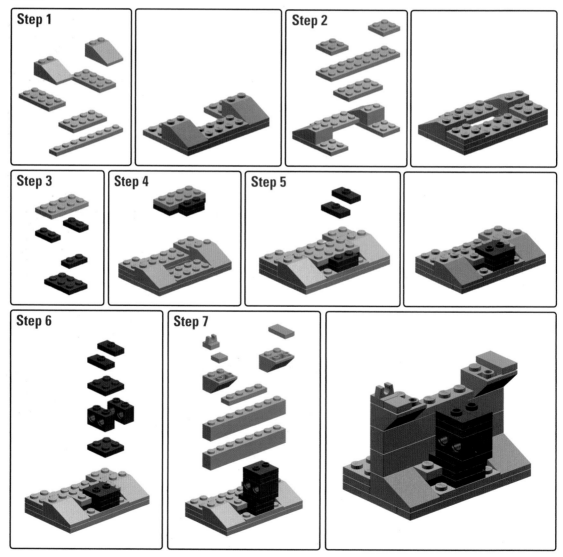

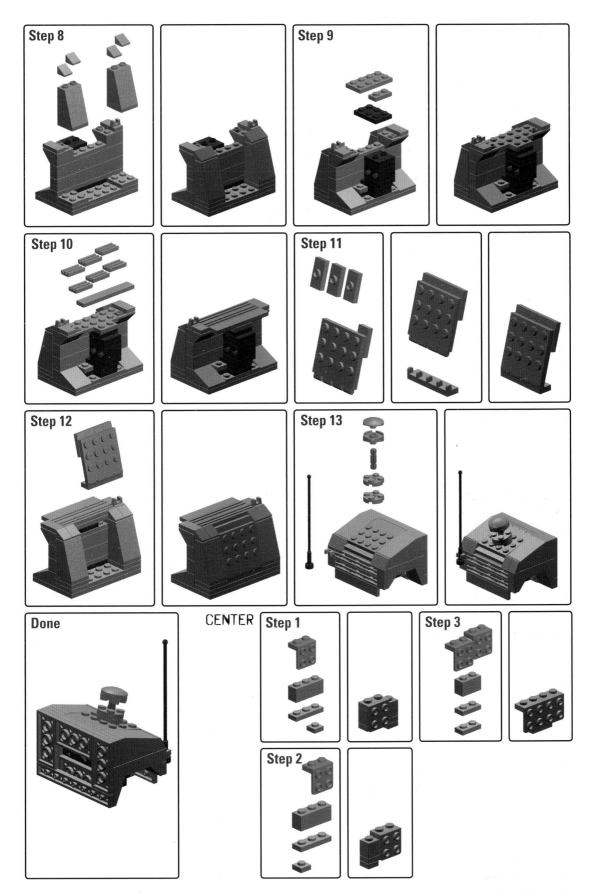

Step 8

Step 9

Step 10

Step 11

Step 12

Step 13

Done

CENTER

Step 1

Step 2

Step 3

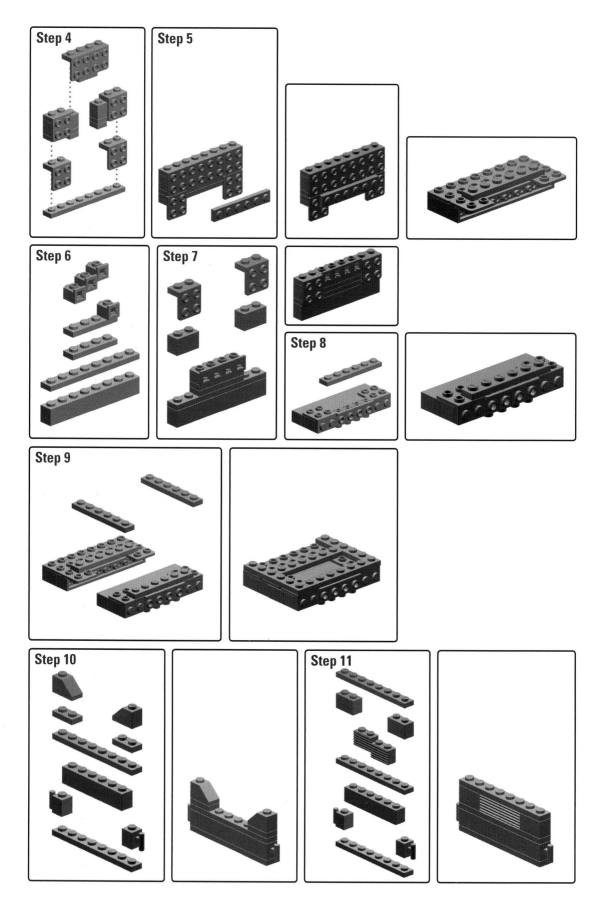

Step 12

Done

Connect the three sections.

Done

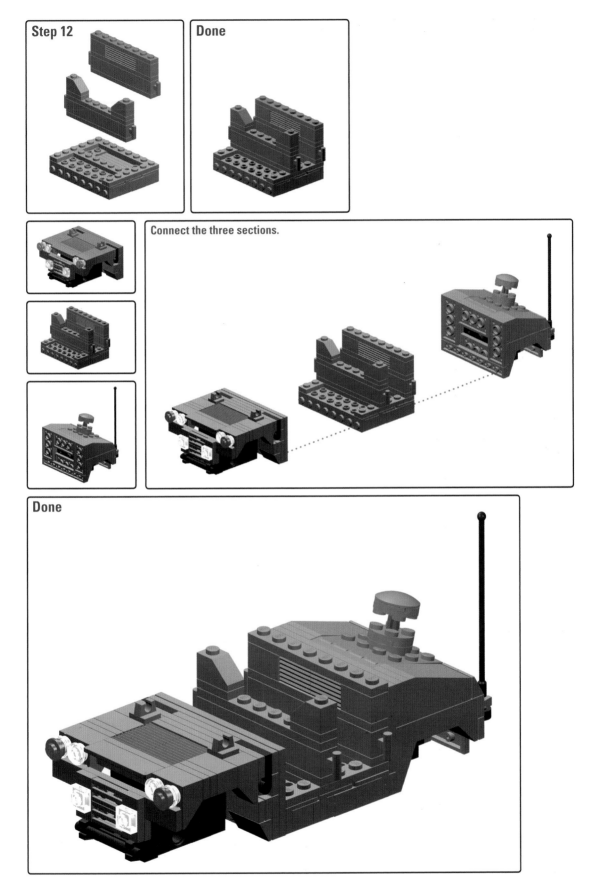

ROOF

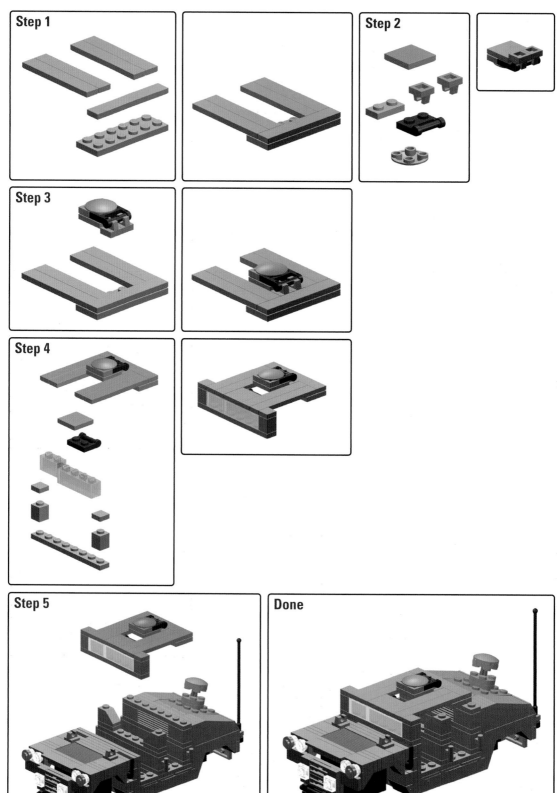

Step 1

Step 2

Step 3

Step 4

Step 5

Done

DOORS

Step 1

x2

Step 2

x2

Step 3

Step 4

Done

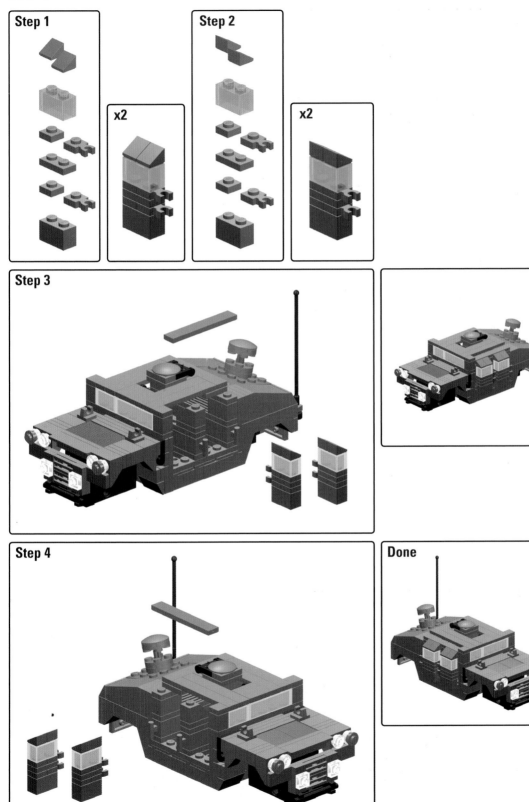

WHEELS AND FINAL ATTACHMENTS

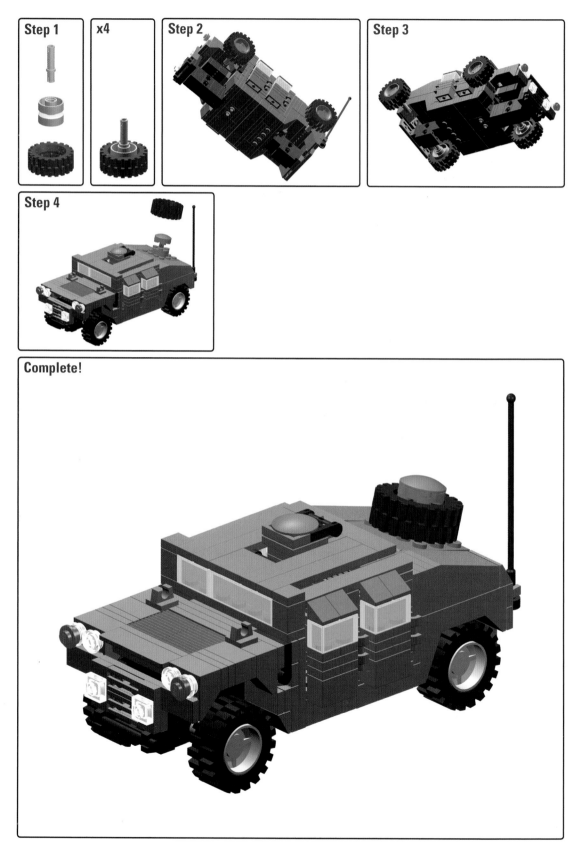

Step 1

x4

Step 2

Step 3

Step 4

Complete!

V-2 MISSLE

SCALE: MINI-FIGURE

The pioneer of all rockets, the V-2 Missile was the first man-made artifact to ever reach outer space. A 1944 German development for use in WWII, this was a long-range ballistic rocket that became the catalyst to modern rockets, the likes of which were used in both the United States and Soviet–era space programs. This powerful forerunner of space exploration was also responsible for the first photographs of the Earth from outside our atmosphere.

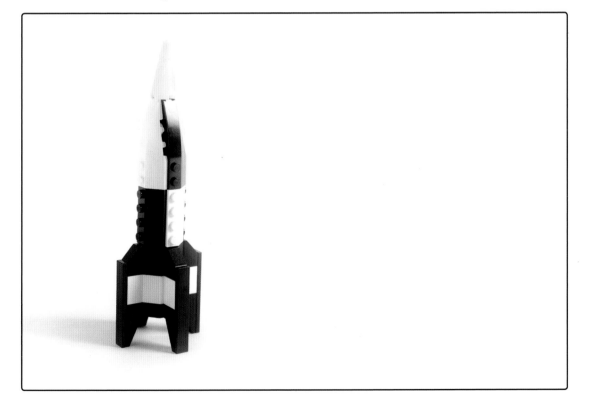

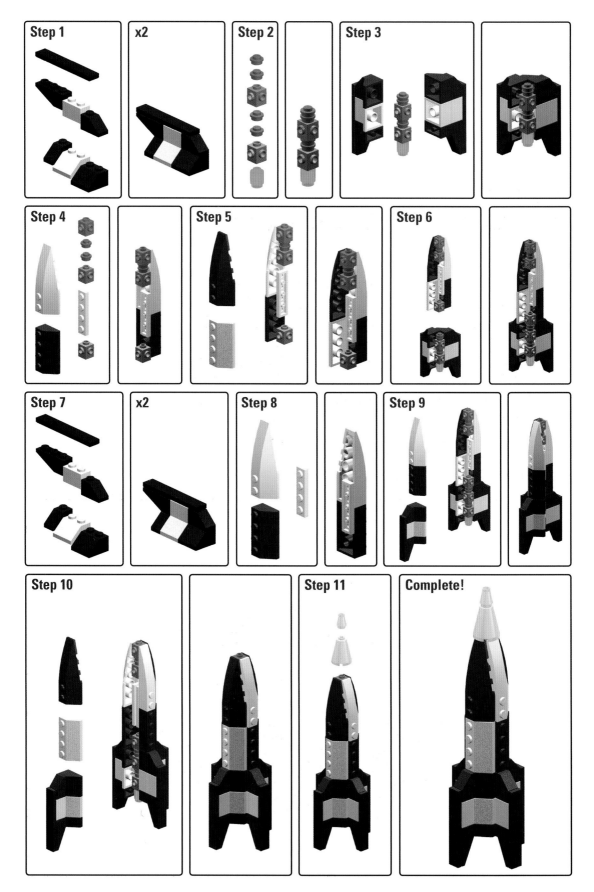

Step 1

x2

Step 2

Step 3

Step 4

Step 5

Step 6

Step 7

x2

Step 8

Step 9

Step 10

Step 11

Complete!

AAVP 7A1
ASSAULT AMPHIBIOUS VEHICLE
SCALE: 1:44

Referred to as Amtracks, or "amphibious tractors,"
by Marines, this modern amphibious vehicle has
been in use since 1984 by NATO forces. Designed
to support troops switching from naval to land
operations, it sports eight smoke grenade tubes, a
.50 M2HB machine gun, and a Mark XIX automatic
grenade launcher. Later models have Mine Clearing
Line Charges (MCLCs) used to clear paths through
mine fields. In the mini-figure LEGO model, the
turret rotates, the bay door opens, and the roof can
be removed to access the interior.

BASE

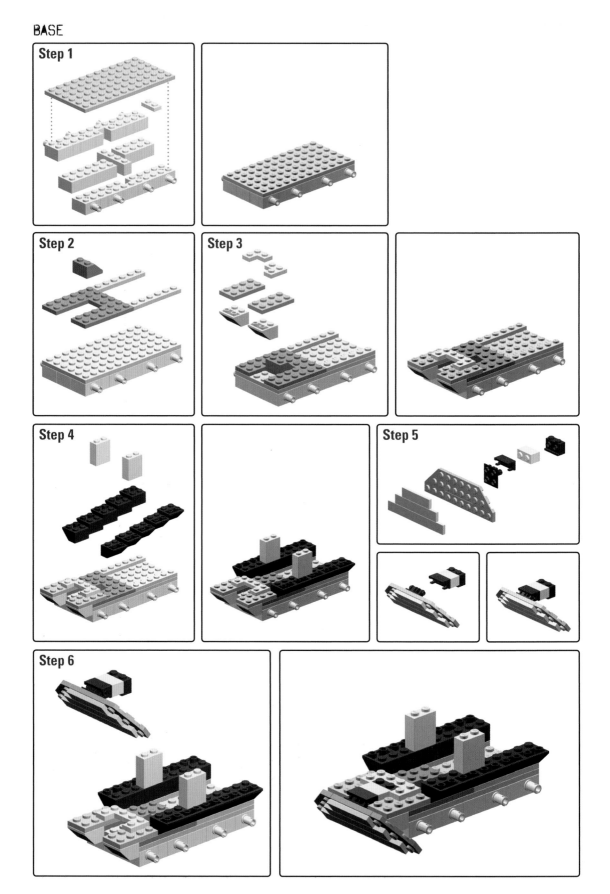

Step 1

Step 2

Step 3

Step 4

Step 5

Step 6

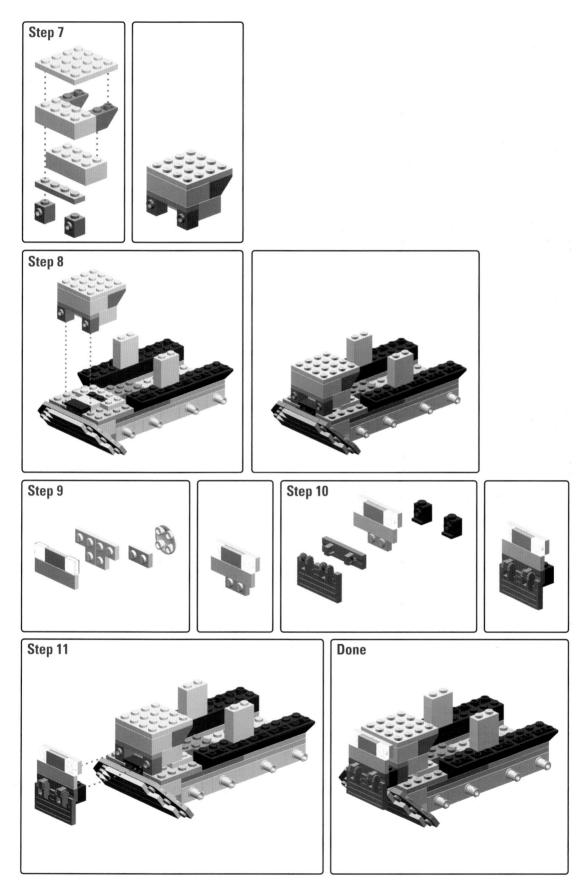

Step 7

Step 8

Step 9

Step 10

Step 11

Done

SIDE PANELS

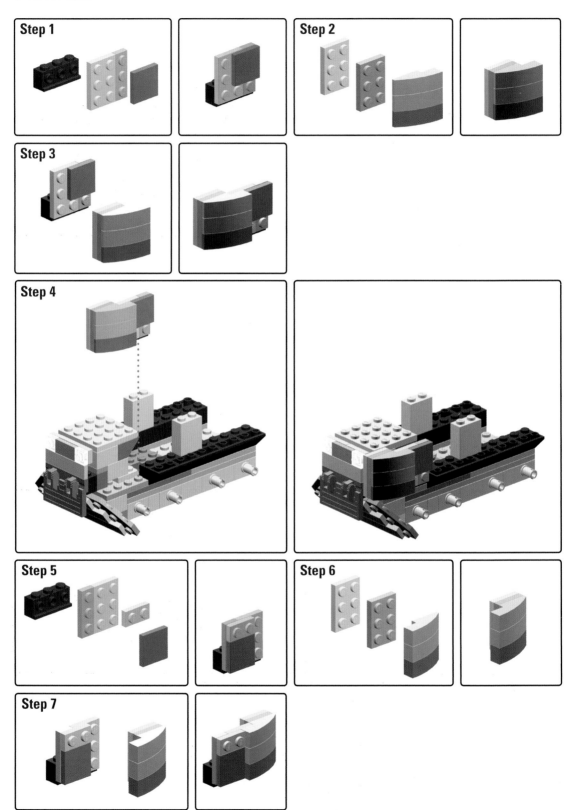

Step 1

Step 2

Step 3

Step 4

Step 5

Step 6

Step 7

Step 8

Step 9

Step 10

Step 11

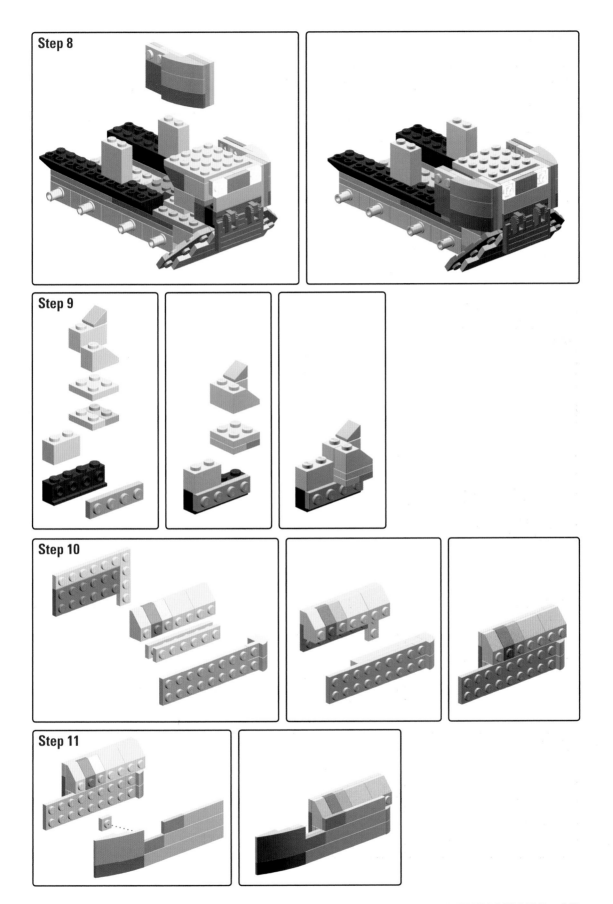

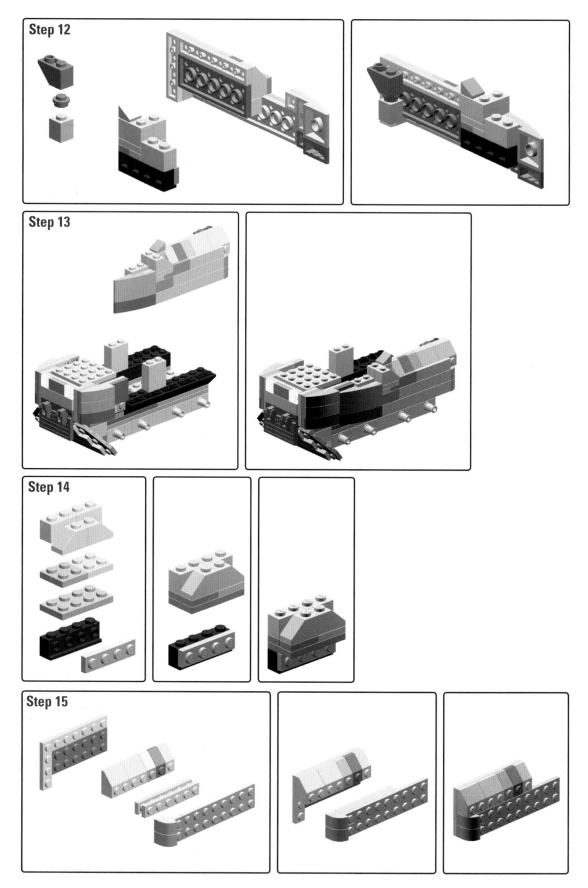

Step 12

Step 13

Step 14

Step 15

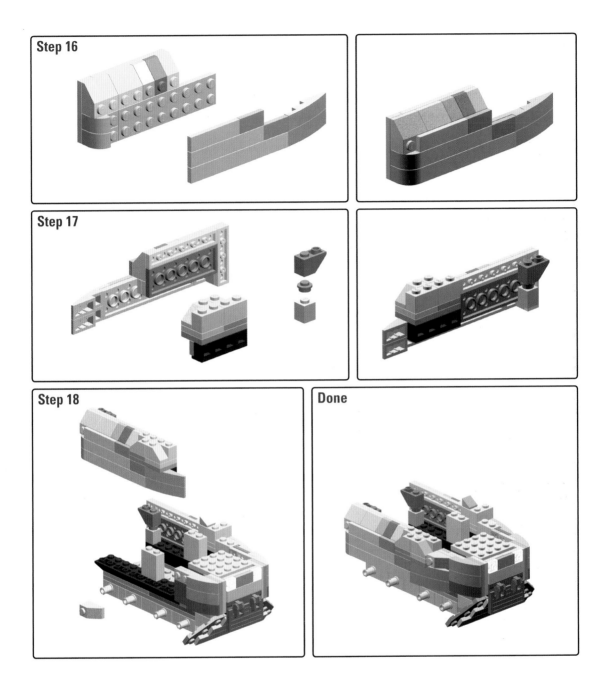

Step 16

Step 17

Step 18

Done

ROOF

Step 1

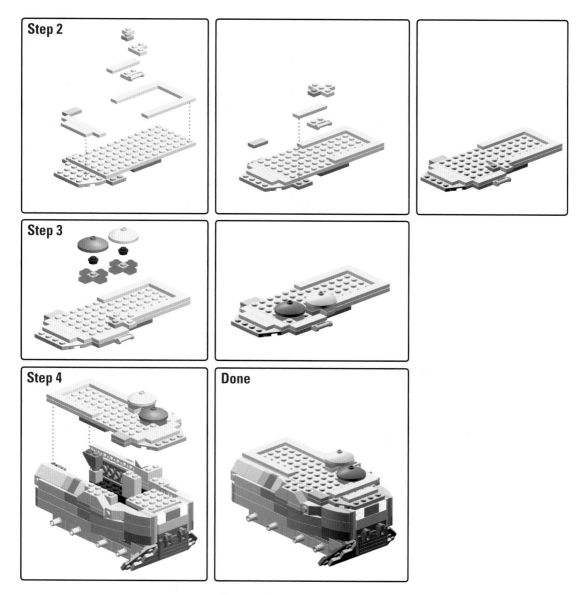

Step 2

Step 3

Step 4

Done

FINAL ATTACHMENTS AND TREADS

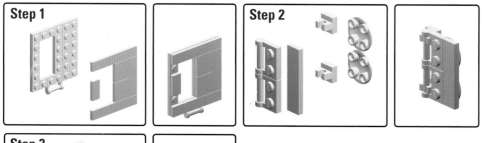

Step 1

Step 2

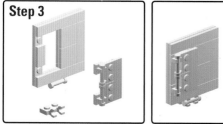

Step 3

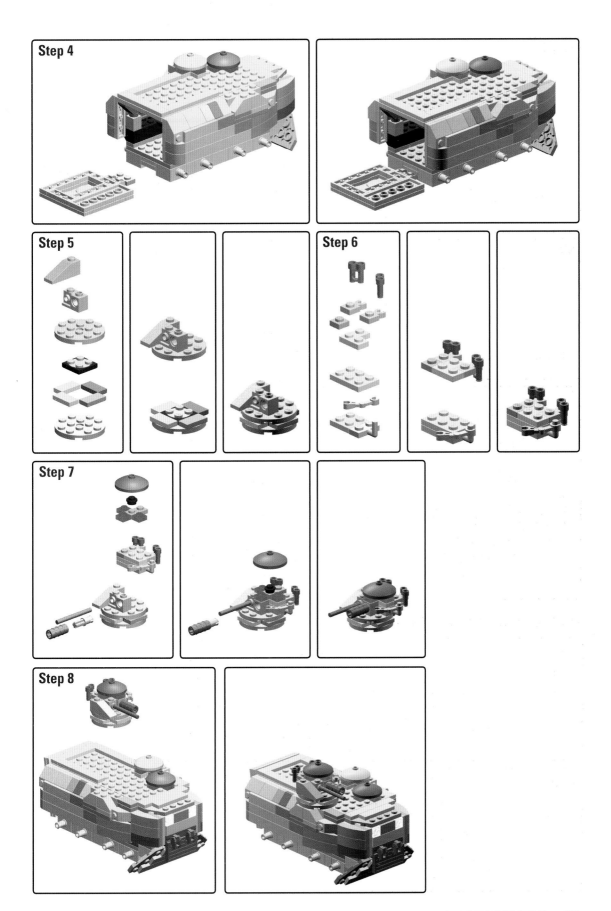

Step 4

Step 5

Step 6

Step 7

Step 8

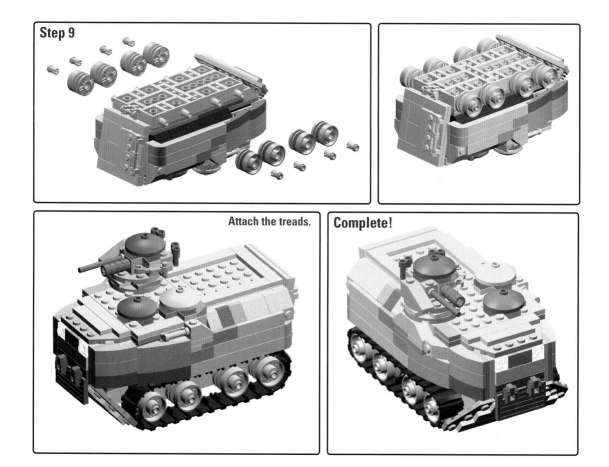

Step 9

Attach the treads.

Complete!

LANDKREUZER P. 1000 BATTLE TANK
SCALE: 1:138

This ultra-heavy tank was designed in 1942 but never made it off the drawing board. This monster of a tank would have weighed 1,000 tons—five times as heavy as any tank ever built—would have carried two naval guns on a rotating triple naval turret resting on a heavy chassis, and would have been armored with ten inches of plate steel. The design required two 8,500 U-boat engines and would have maneuvered easily through water with the help of snorkels. The LEGO construction of the Landkreuzer requires a remote for the electronic functions to work.

MAIN TURRET

Step 1

Step 2

Step 3

Step 4

Step 5

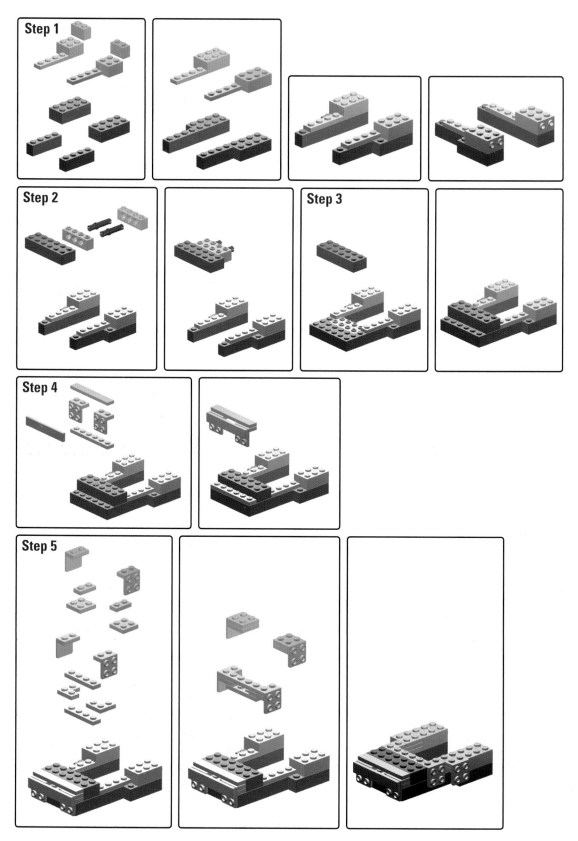

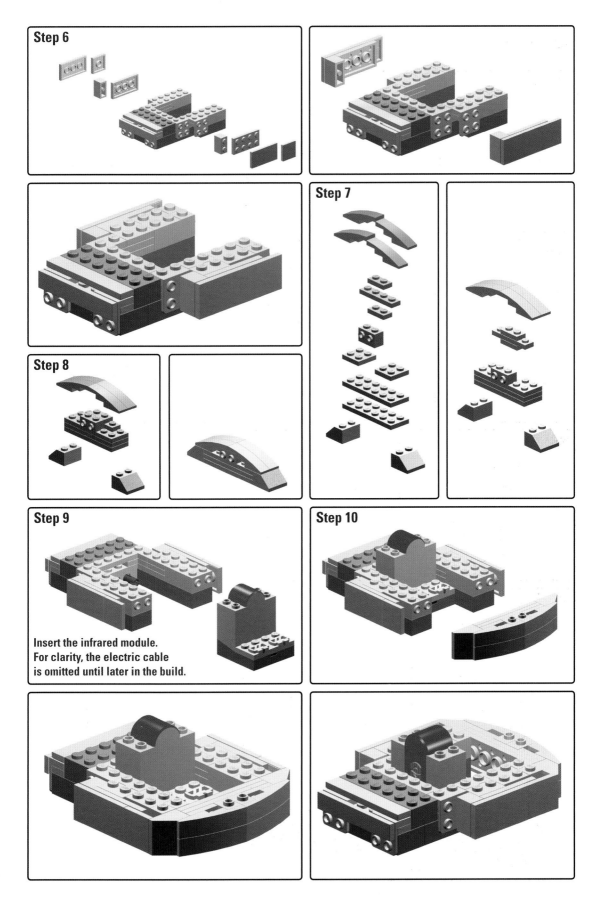

Step 6

Step 7

Step 8

Step 9

Insert the infrared module.
For clarity, the electric cable
is omitted until later in the build.

Step 10

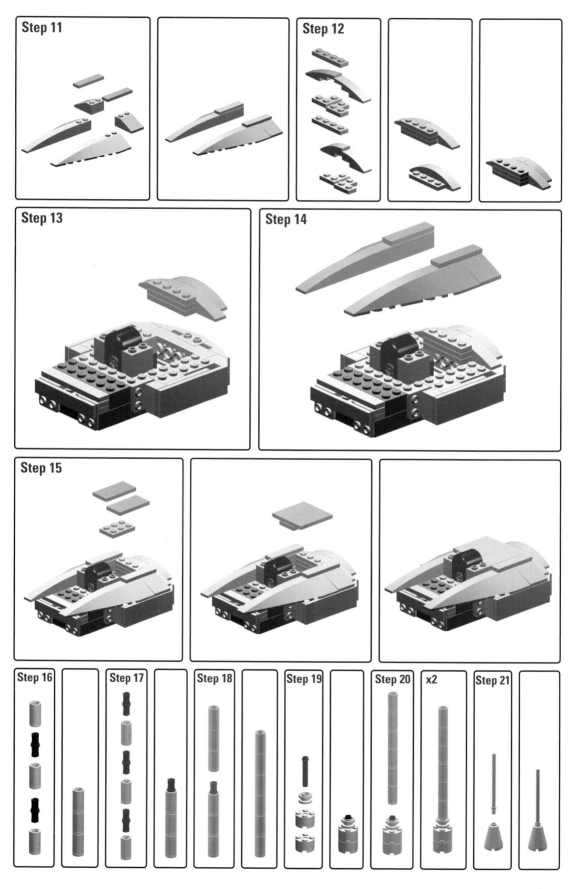

Step 11

Step 12

Step 13

Step 14

Step 15

Step 16

Step 17

Step 18

Step 19

Step 20　x2

Step 21

Step 22

Step 23

Step 24

Step 25

Done

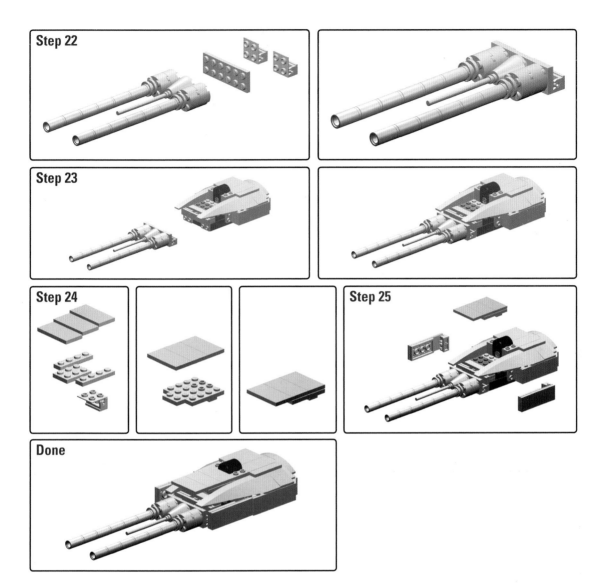

WHEEL ASSEMBLIES

Step 1

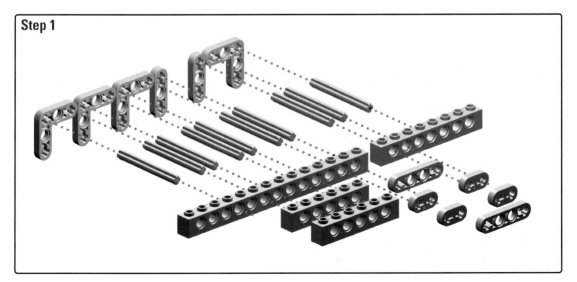

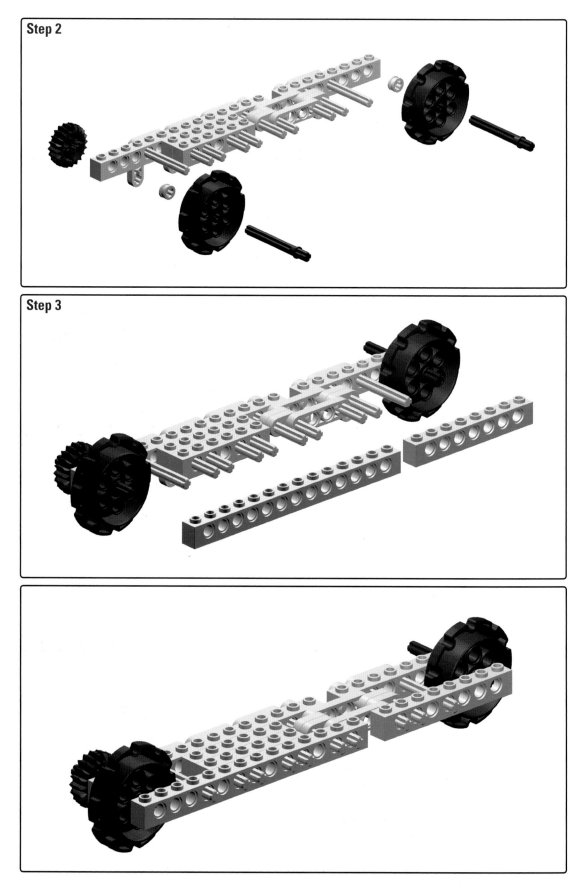

Step 2

Step 3

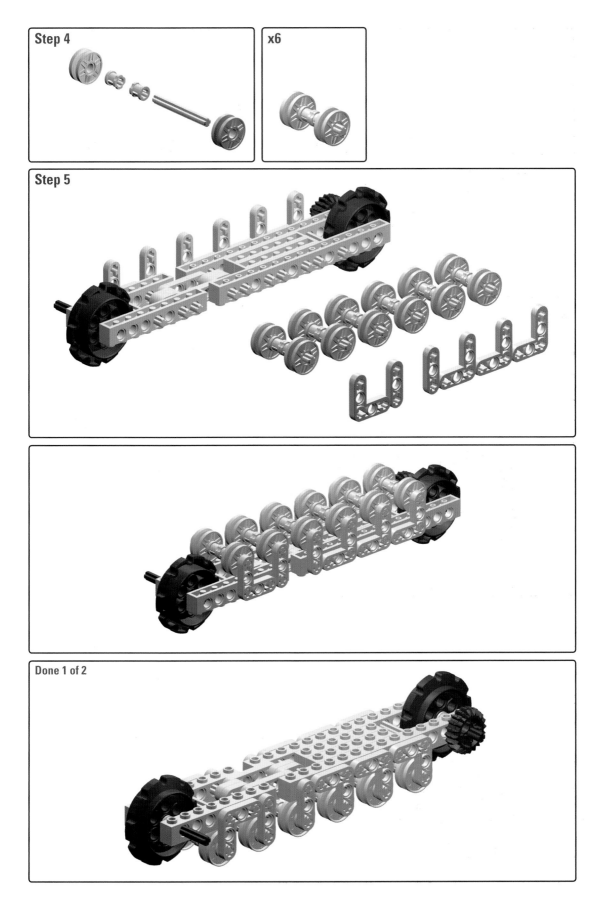

Step 4

x6

Step 5

Done 1 of 2

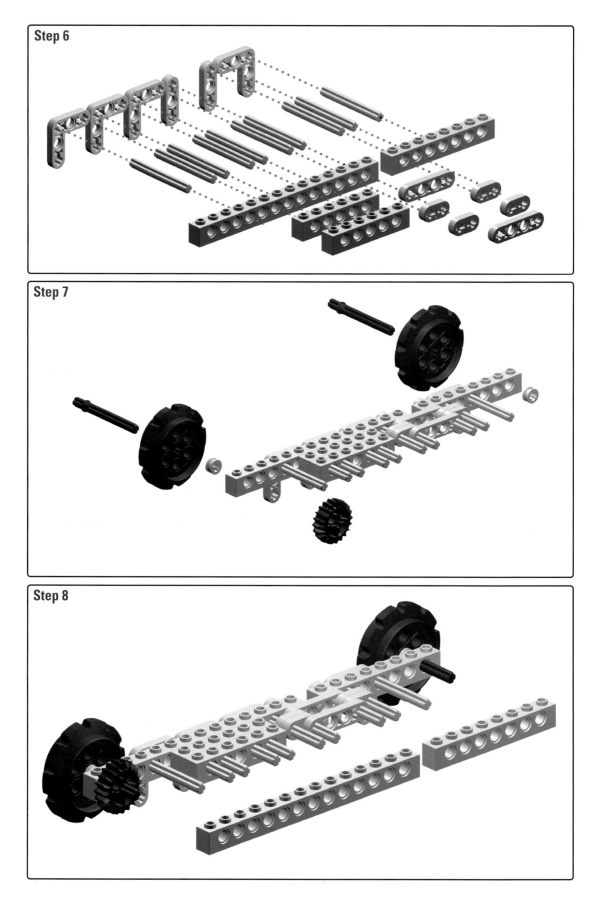

Step 6

Step 7

Step 8

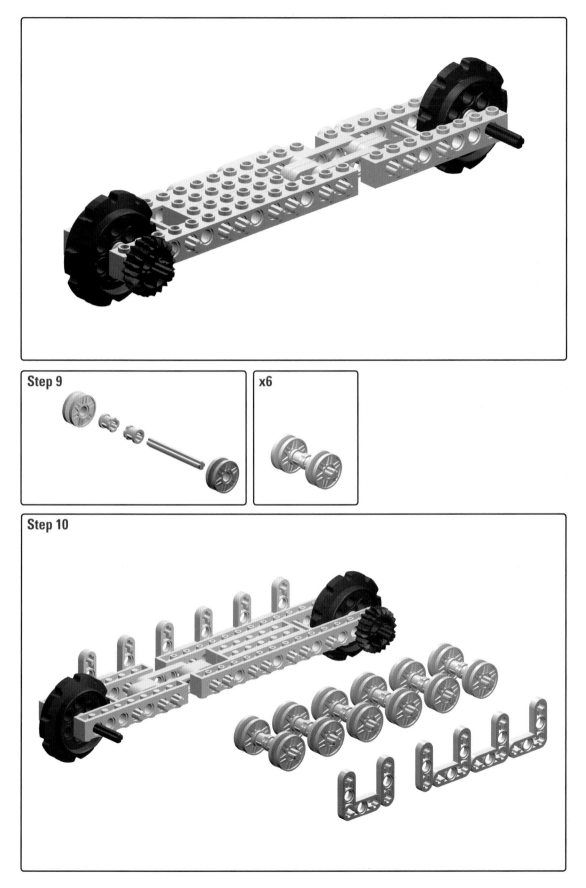

Step 9

x6

Step 10

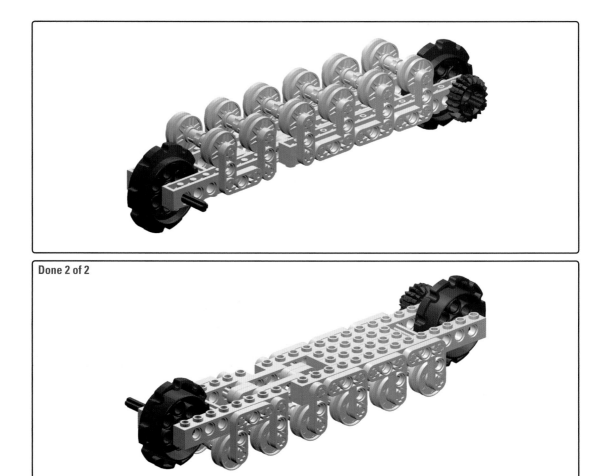

MAIN BODY

Step 1

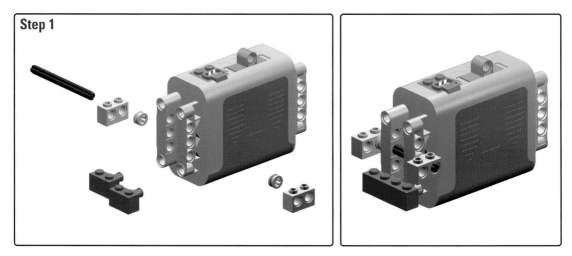

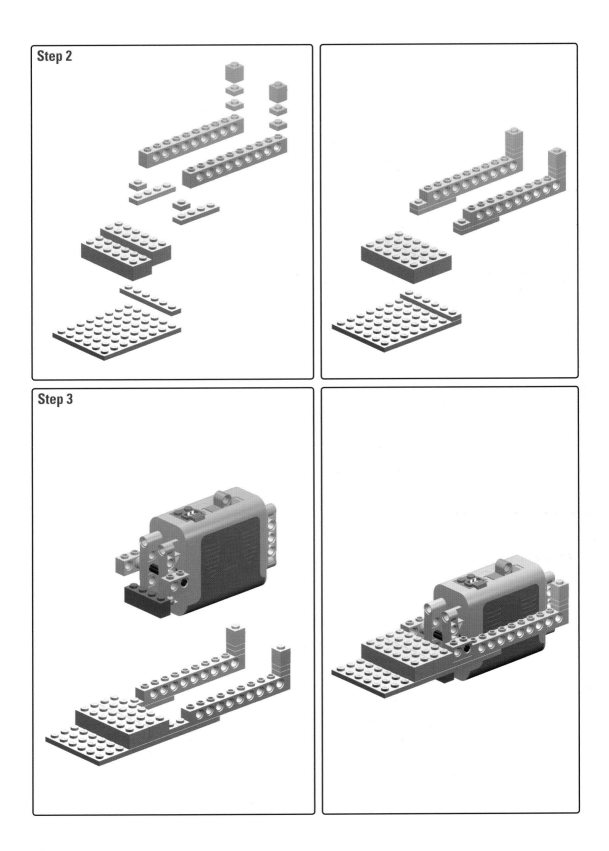

Step 2

Step 3

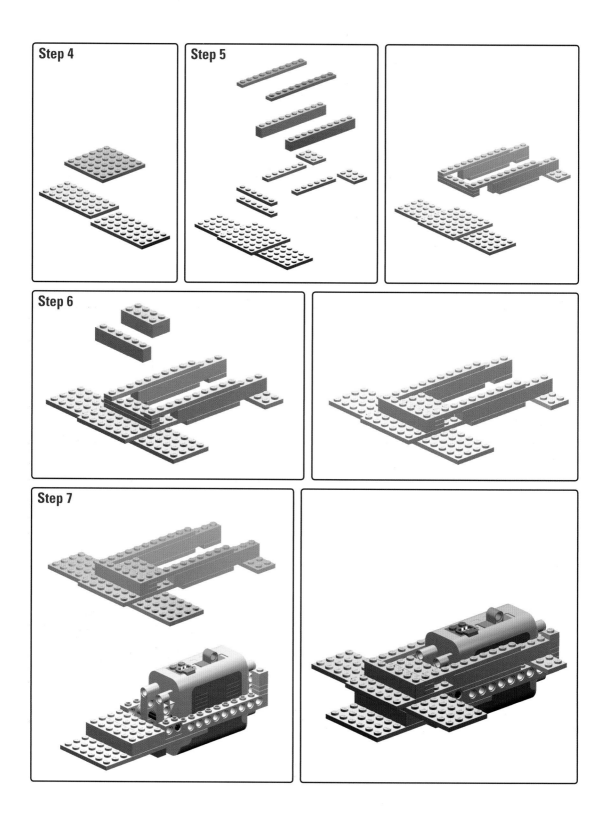

Step 4

Step 5

Step 6

Step 7

Step 8

Step 9

Step 10

Step 11

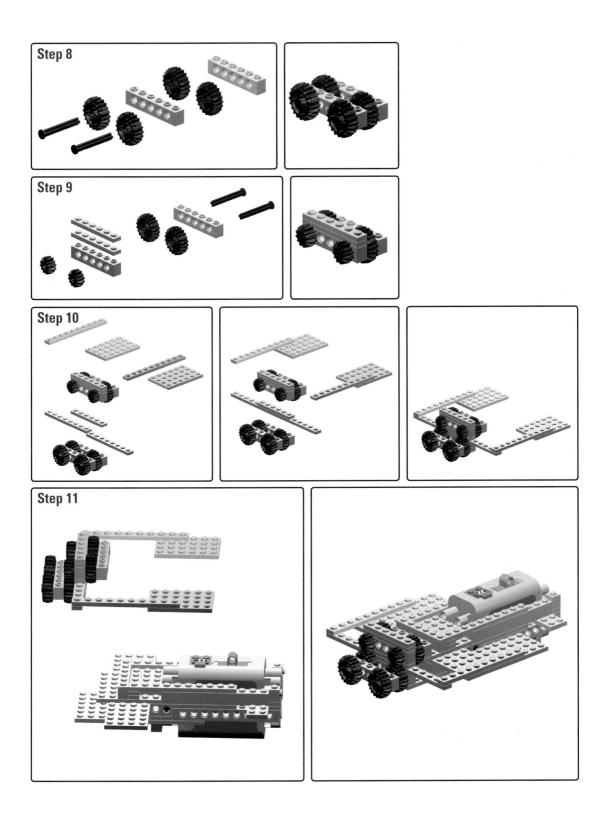

Step 12

Step 13

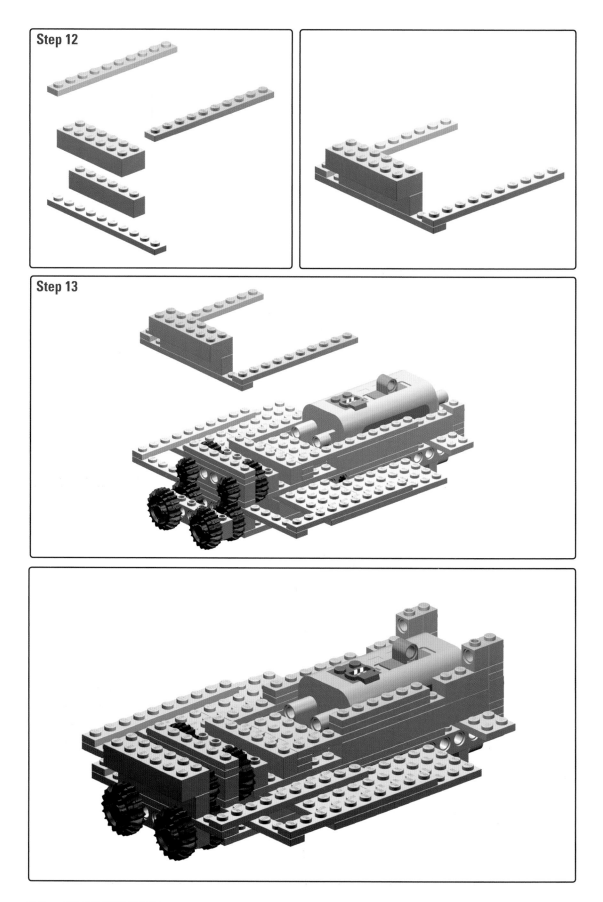

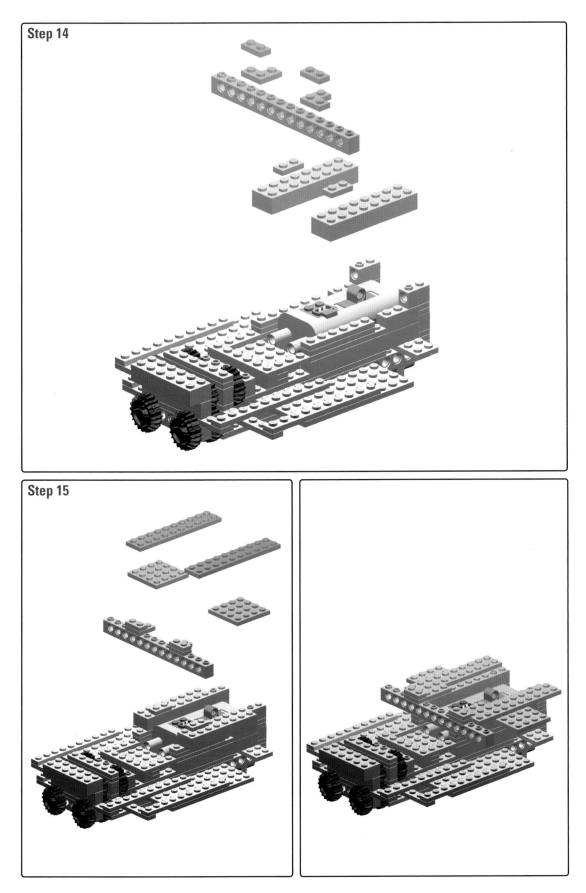

Step 14

Step 15

Step 16

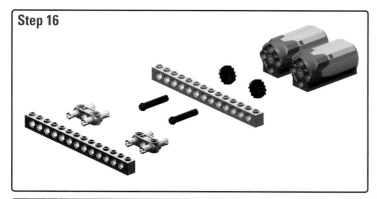

Step 17

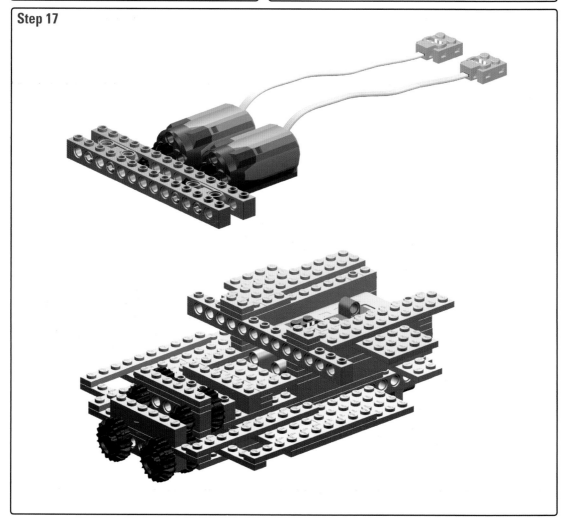

Done

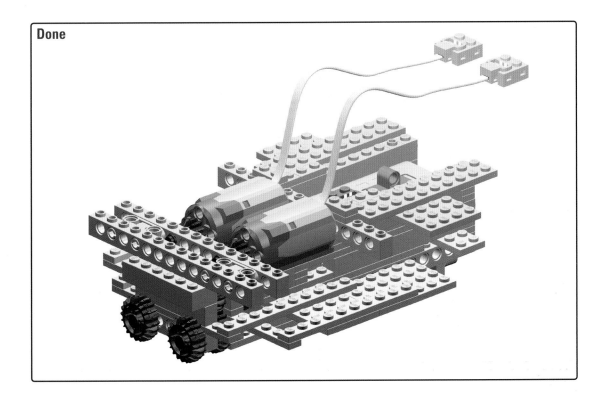

WHEEL ASSEMBLIES AND ARMOR PLATING

Step 1

Remove or loosen the highlighted bricks to attach the wheel assembly.

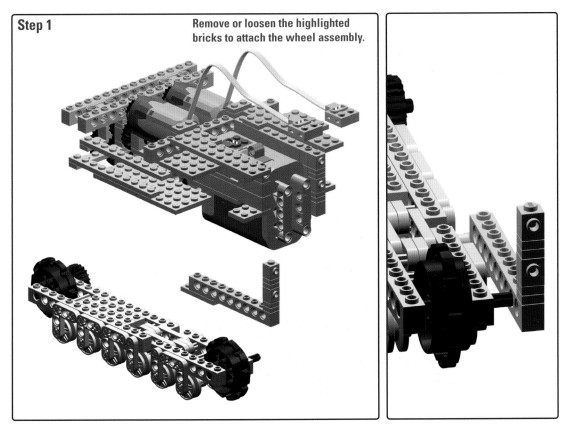

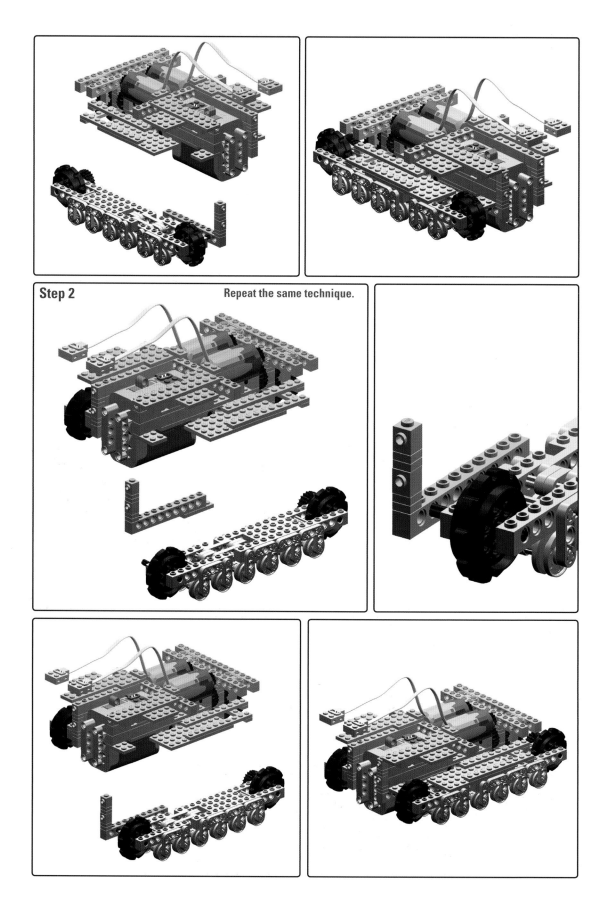

Step 2 Repeat the same technique.

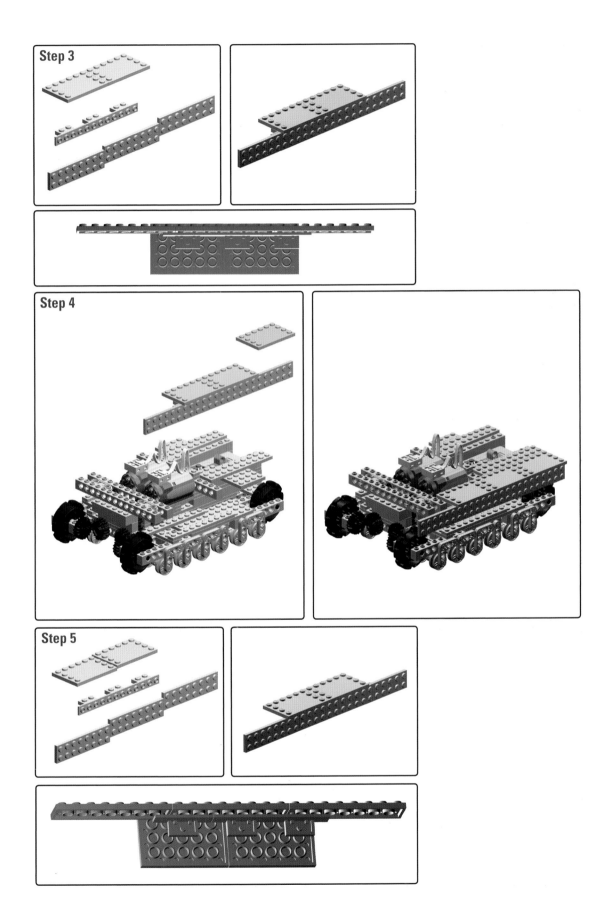

Step 3

Step 4

Step 5

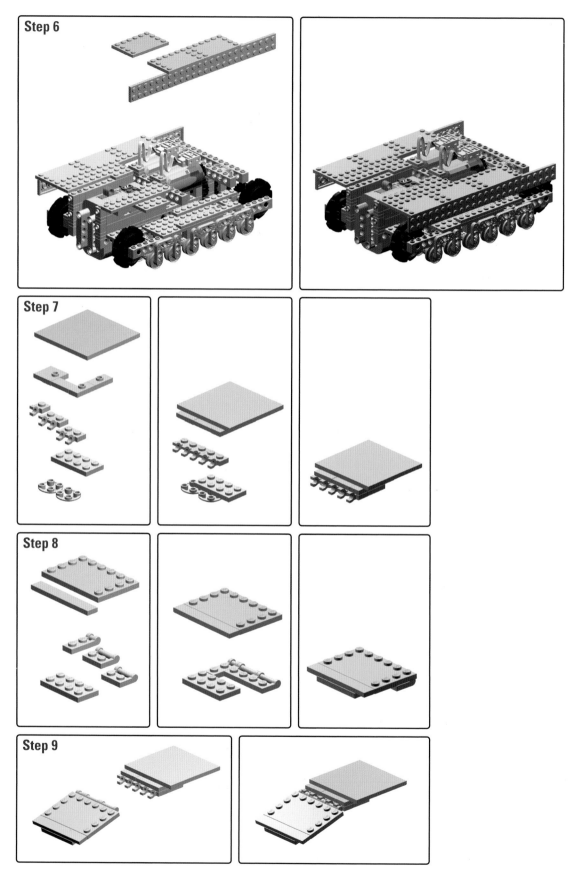

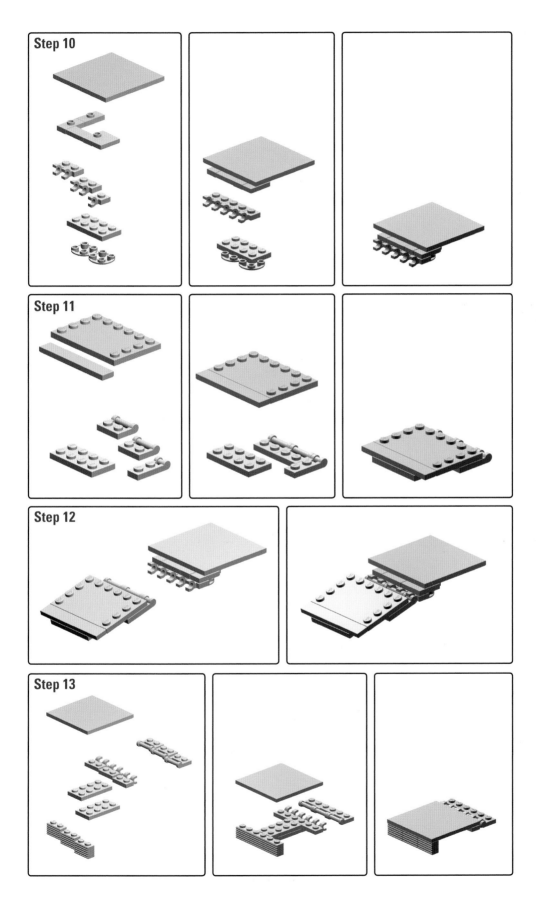

Step 10

Step 11

Step 12

Step 13

Step 14

Step 15

Step 16

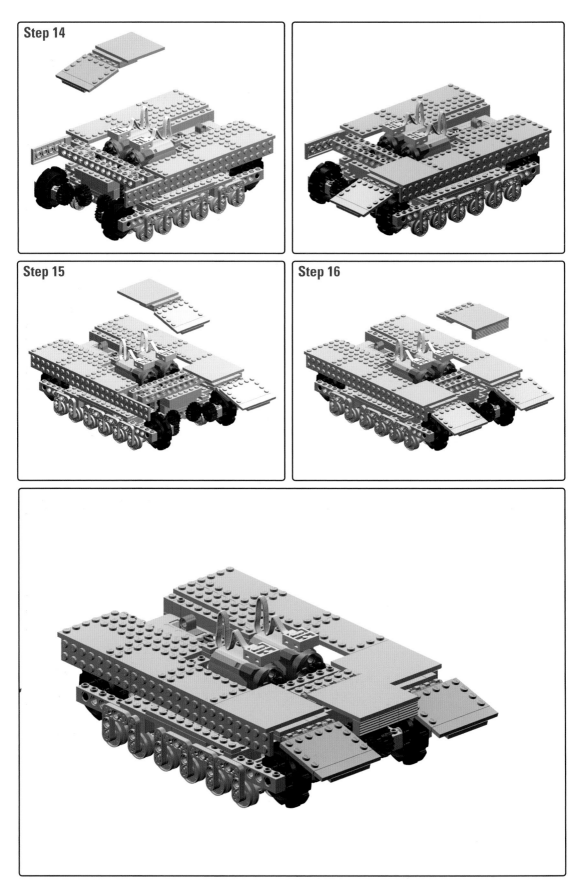

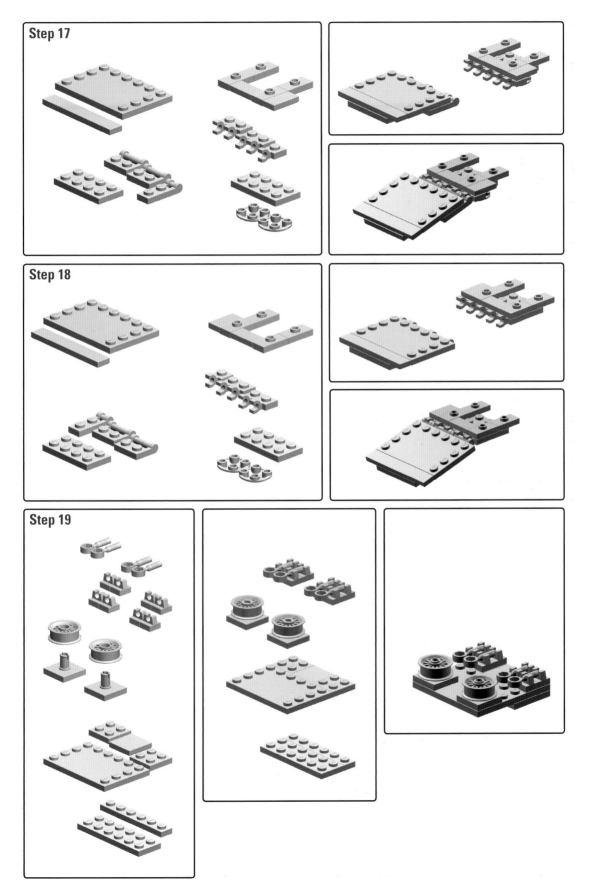

Step 17

Step 18

Step 19

Step 20

Step 21

Step 22

Done

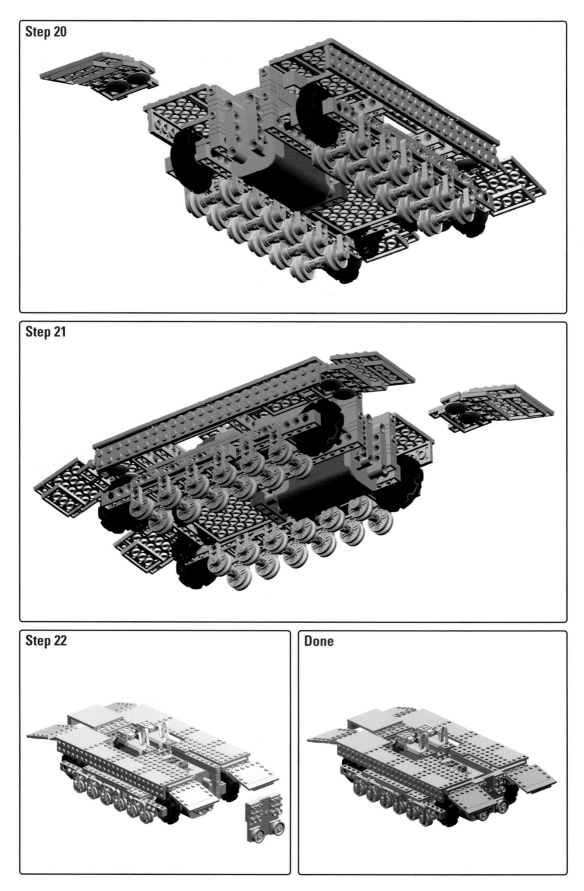

Step 1

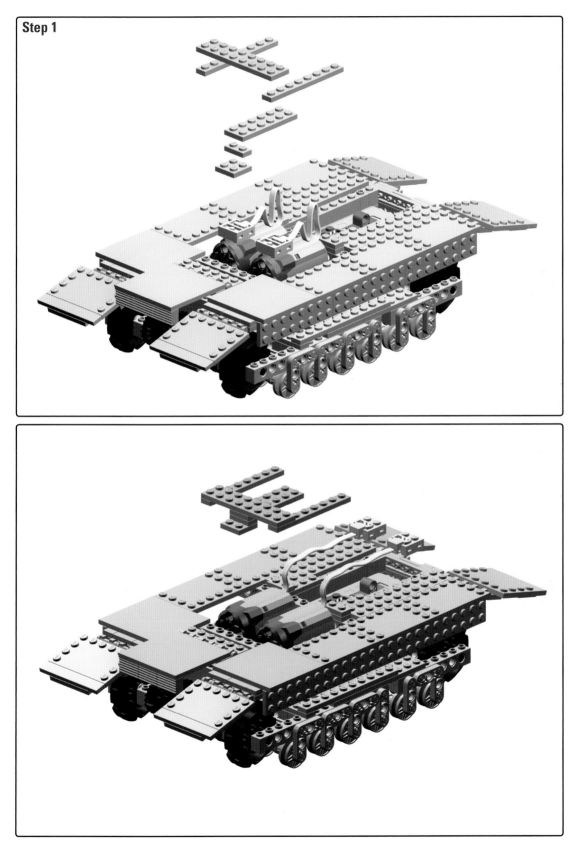

Step 2

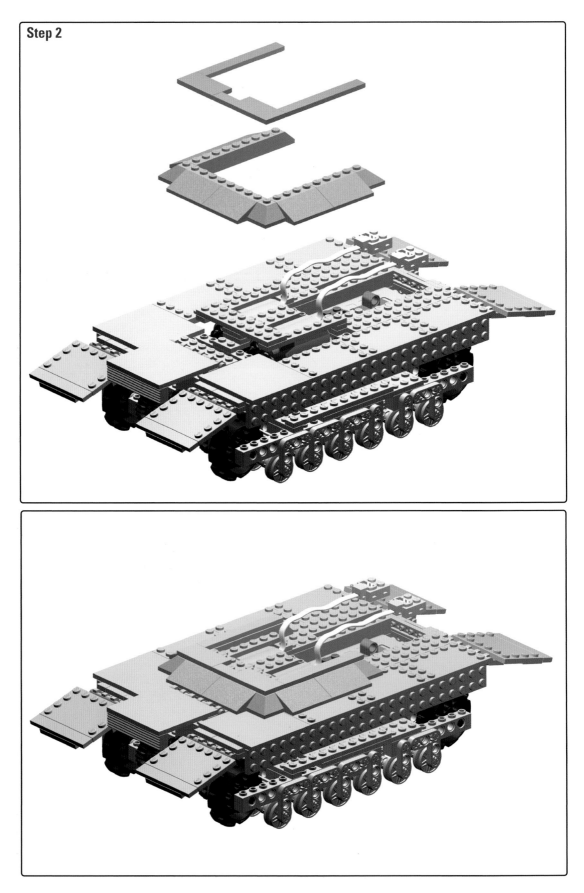

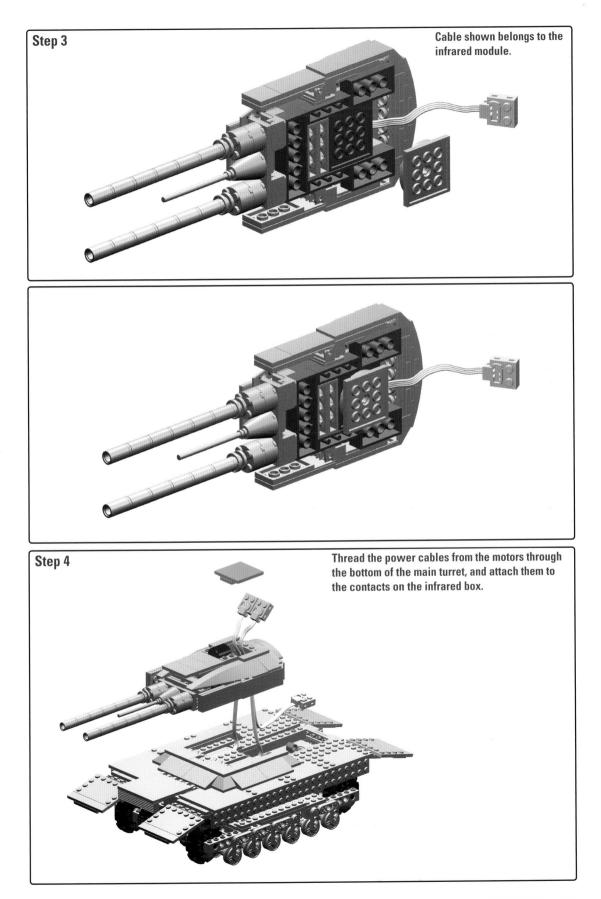

Step 3

Cable shown belongs to the infrared module.

Step 4

Thread the power cables from the motors through the bottom of the main turret, and attach them to the contacts on the infrared box.

Step 5

Step 6

Done

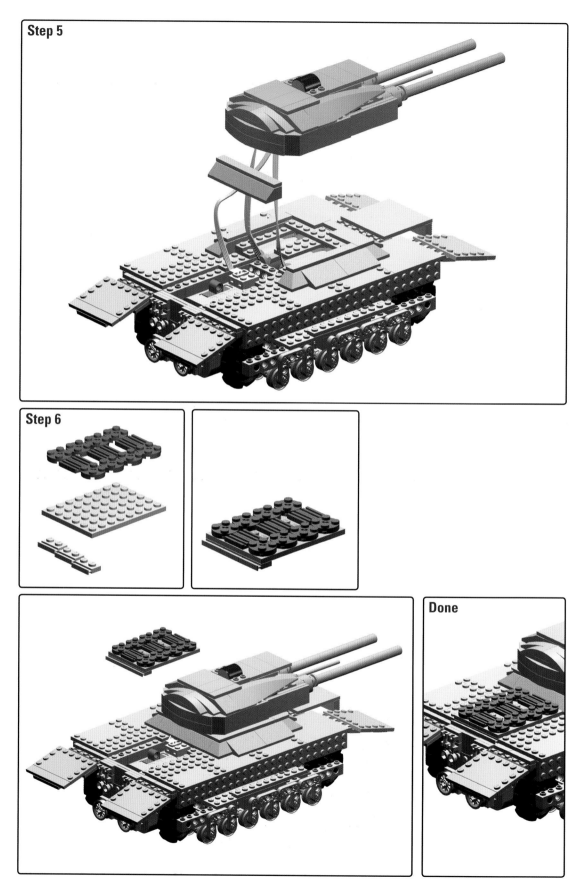

REAR TURRETS, ARMOR, AND TREADS

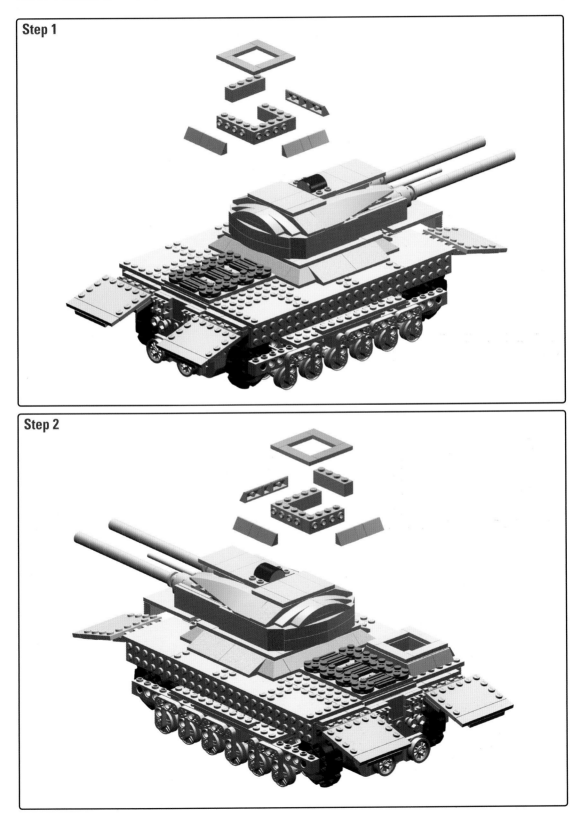

Step 1

Step 2

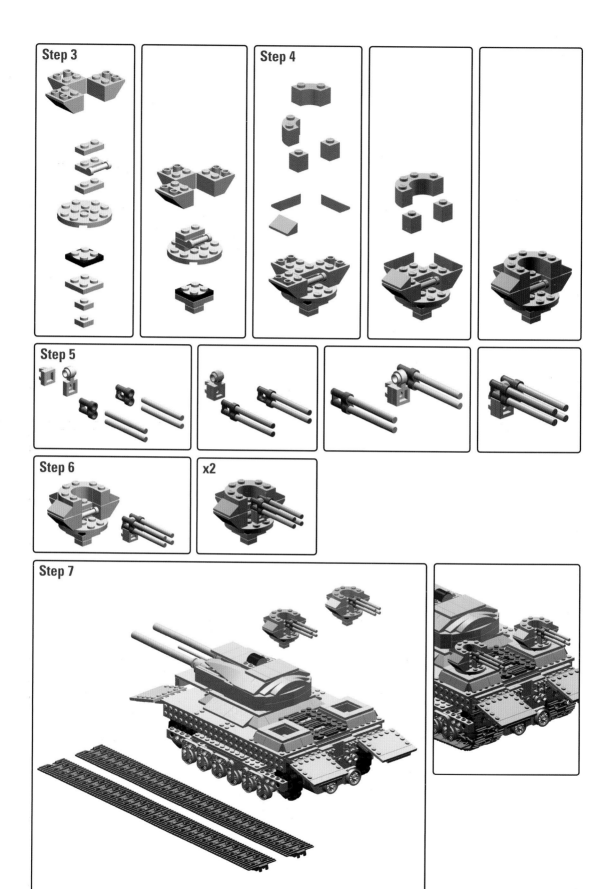

Step 3

Step 4

Step 5

Step 6

x2

Step 7

Step 8

Complete!

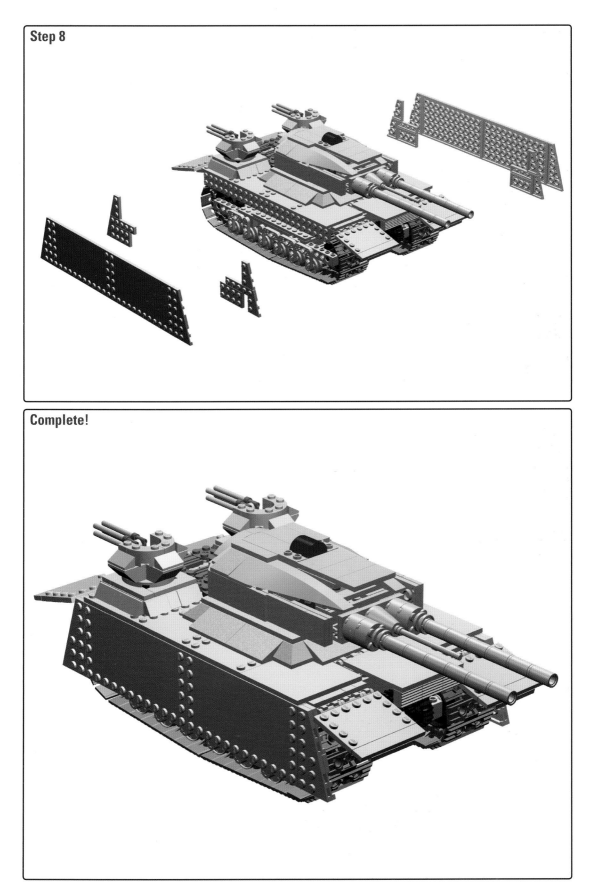

CHAPTER FOUR
SPECIAL WEAPONS

LOS ANGELES-CLASS US SUB
SCALE: 1:434

This modern submarine, built circa 1972, was
designed to stay under water for a long time—
up to a year or more. It is nuclear-powered and
can launch Tomahawk Cruise Missiles. There
are several Los Angeles-class subs still used
by the US Navy. These undersea boats are very
fast and can dive very deep—the exact specs
are still classified.

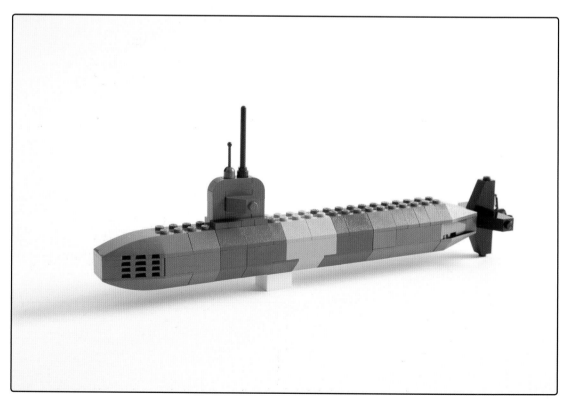

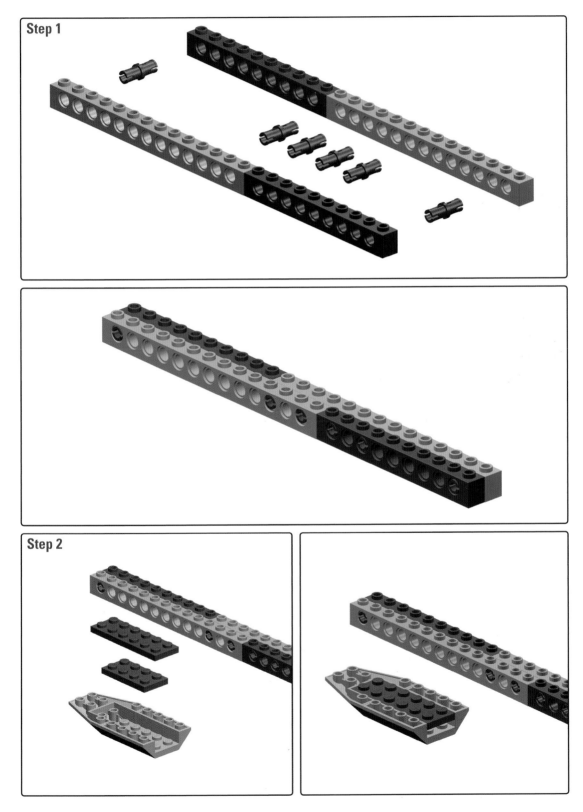

Step 1

Step 2

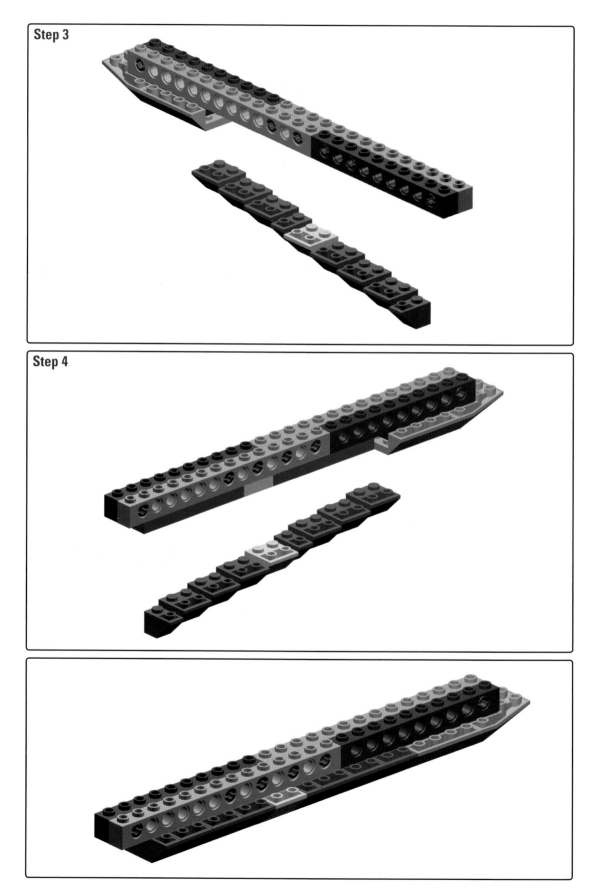

Step 3

Step 4

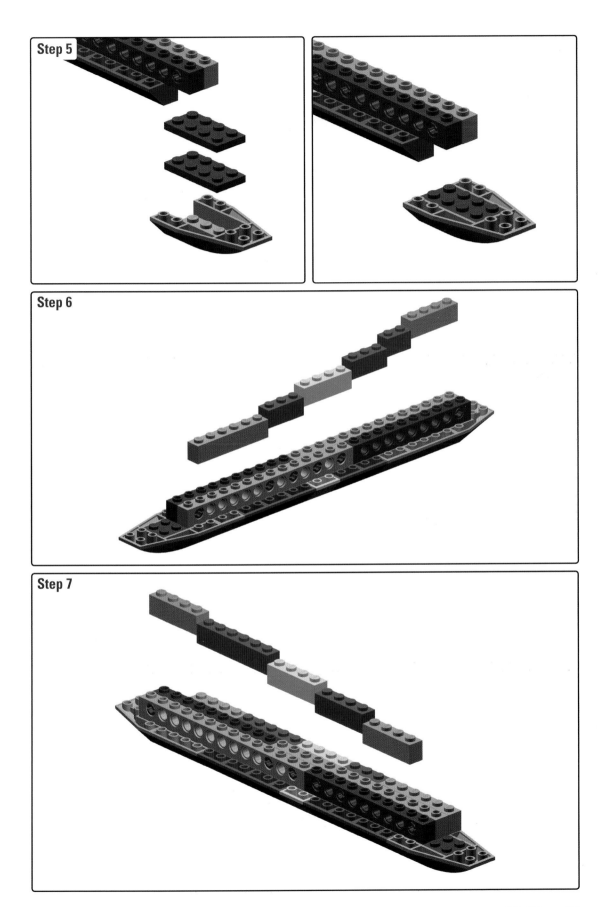

Step 5

Step 6

Step 7

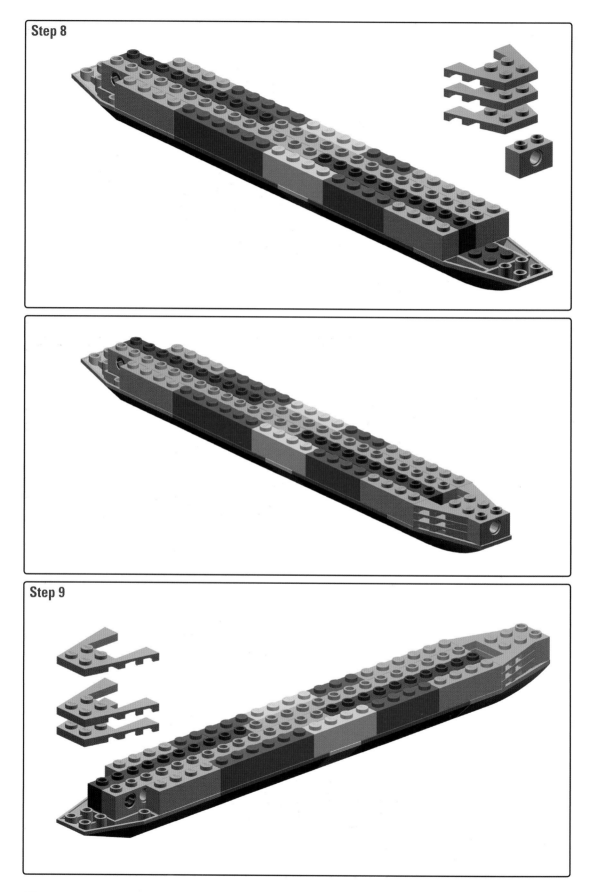

Step 8

Step 9

Step 10

Step 11

Step 12

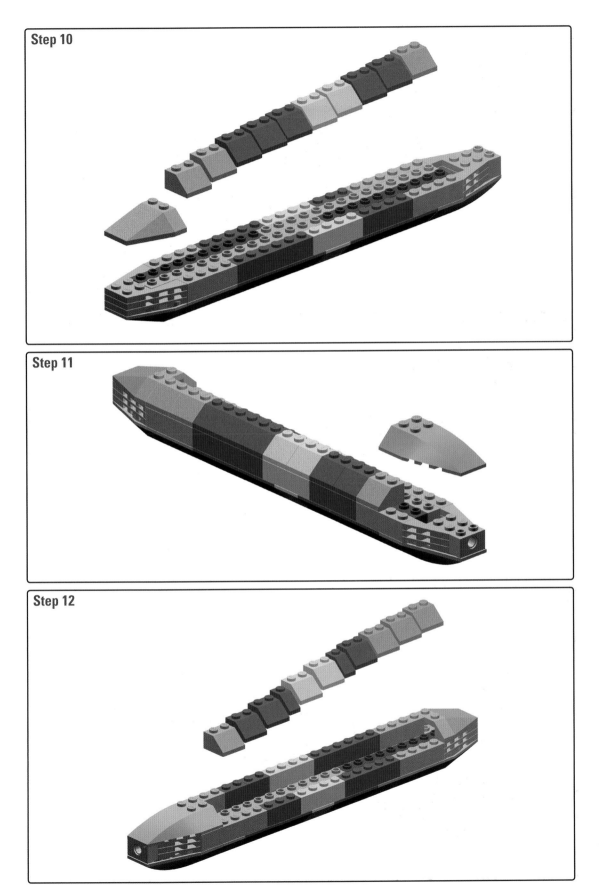

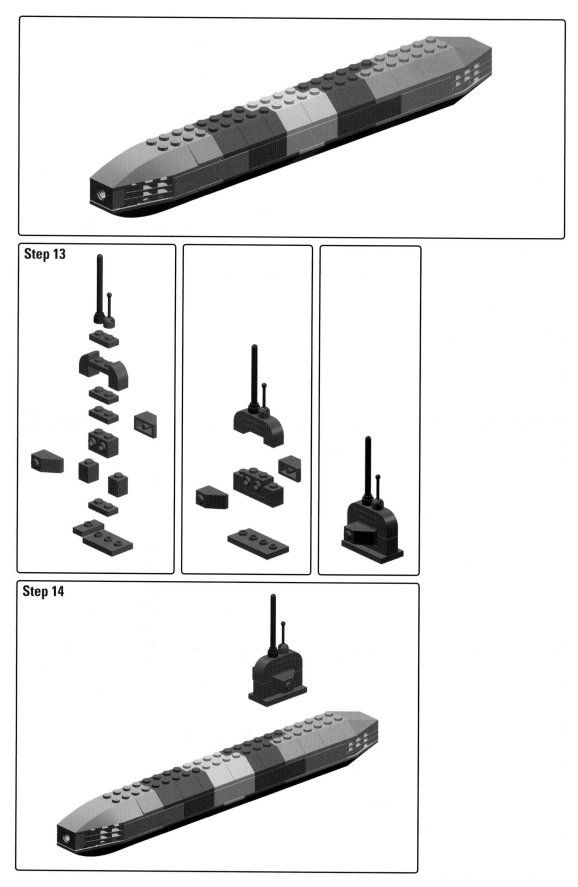

Step 13

Step 14

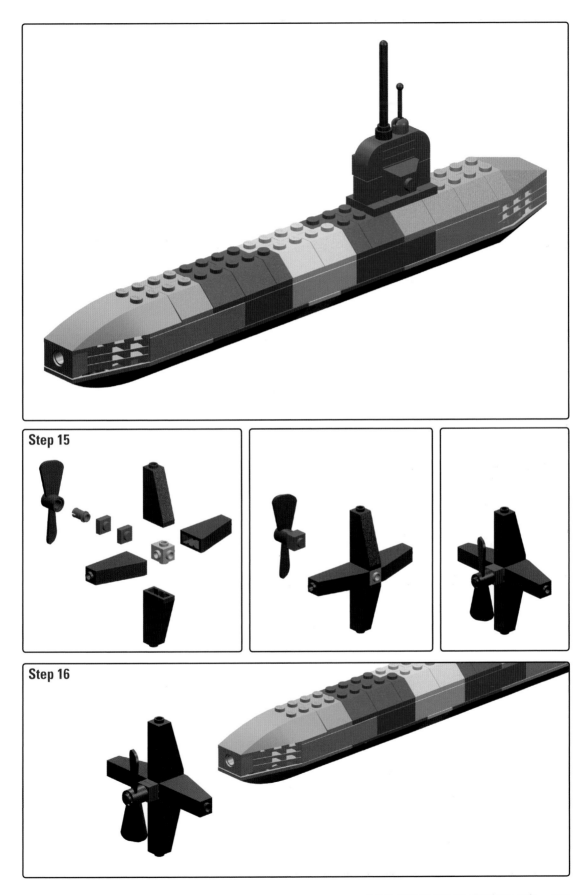

Step 15

Step 16

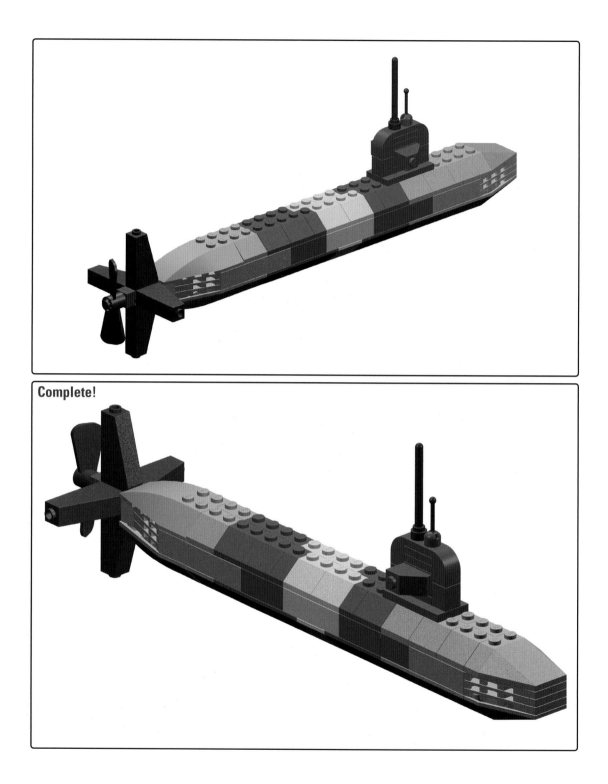

Complete!

GOLIATH TRACKED MINE
SCALE: 1:20

The *Leichter Ladungsträger Goliath* was used as a remote-control demolition vehicle. Only four feet long and about a foot tall, the caterpillar wheels made it useful on many terrains, and it could take down bridges, buildings, tanks, and soldiers by carrying explosives. A good idea in theory but not in form, the Goliath Tracked Mine had to be connected to the remote with a wire. Allied troops often cut or destroyed the wire, rendering the tank completely useless. Early models had an electric engine, but for cost reasons, small gas engines were added. They were used early on in the D-Day invasion until they met heavy gunfire and explosions.

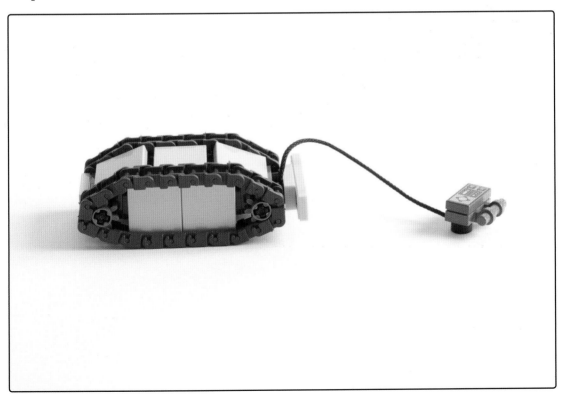

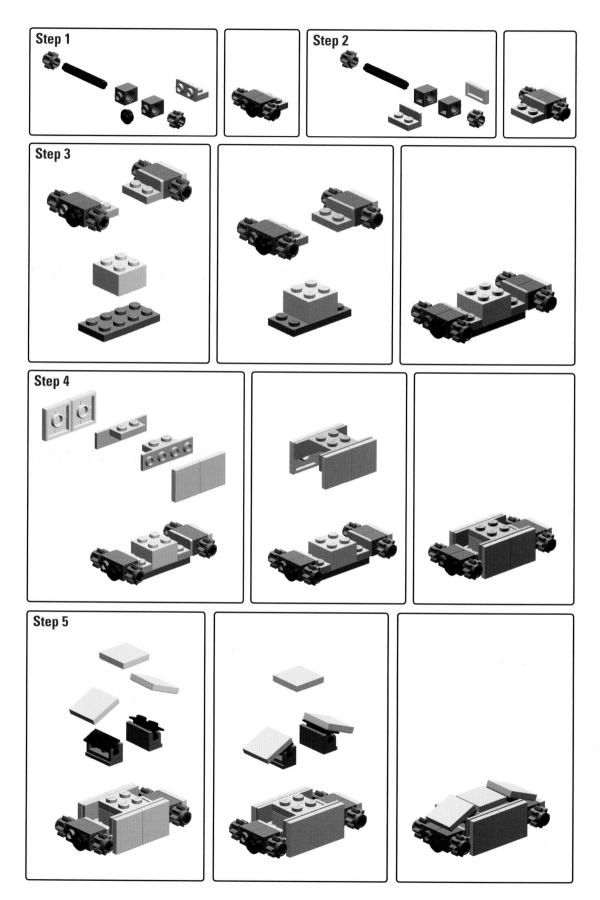

Step 1

Step 2

Step 3

Step 4

Step 5

Step 6

Step 7

Complete!

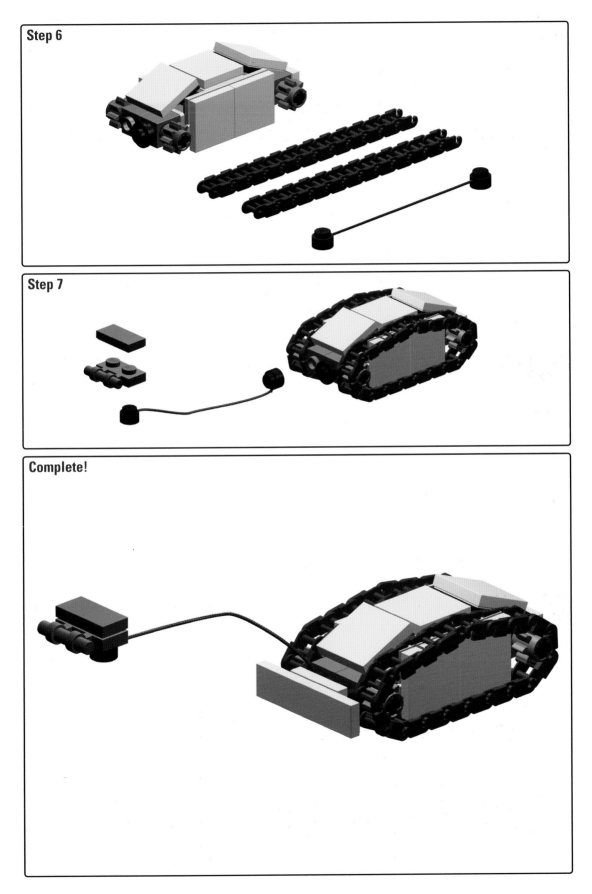

FLAMETHROWER
SCALE: MINI-FIGURE

There is nothing fiercer than harnessing the power of fire. This modern fire-breathing device was first produced in 1901 as the German *flammenwerfer*, but the concept has been around since the Byzantine Empire. With a backpack-like tank and a nozzle to shoot flames, the flamethrowers used in WWI and WWII often spouted flammable liquid, while safer manifestations for agricultural use generally spray propane or gas. Early flamethrowers only shot about sixty feet, but modern devices—though no longer permitted by the military—can reach up to 270 feet.

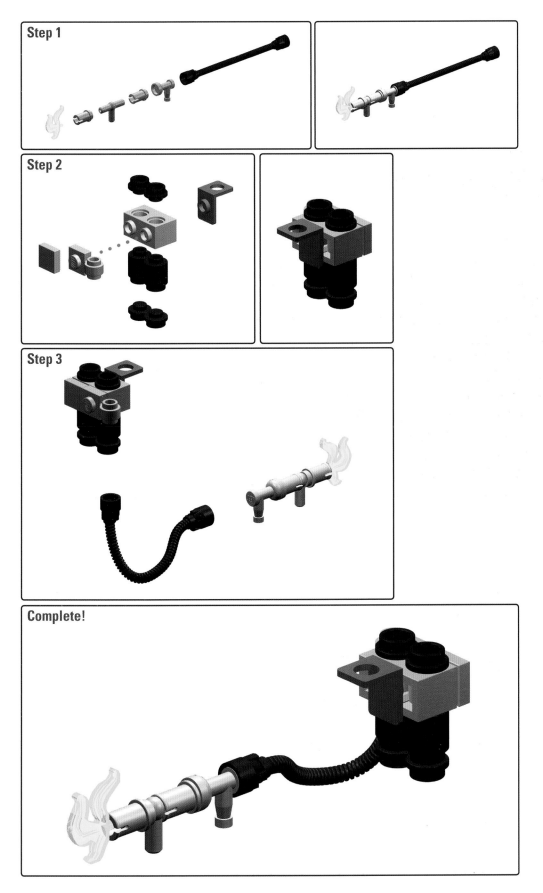

Step 1

Step 2

Step 3

Complete!

GUILLOTINE
SCALE: 1:44

Killer of all killers, the infamous guillotine took no prisoners. This execution device was invented to make beheadings easier and more humane. The device relied on gravity to bring its sharp blade plummeting to the base of the structure where the victim's neck resided, but many unlucky victims encountered a dull edge. Guillotine-like devices were used as early as 1300, but the real deal was popularized during the 1790s, where it was the primary mode of execution during the French Revolution. During this time, the guillotine executed somewhere between 16,000 and 40,000 people, including King Louis XVI and Queen Marie Antoinette.

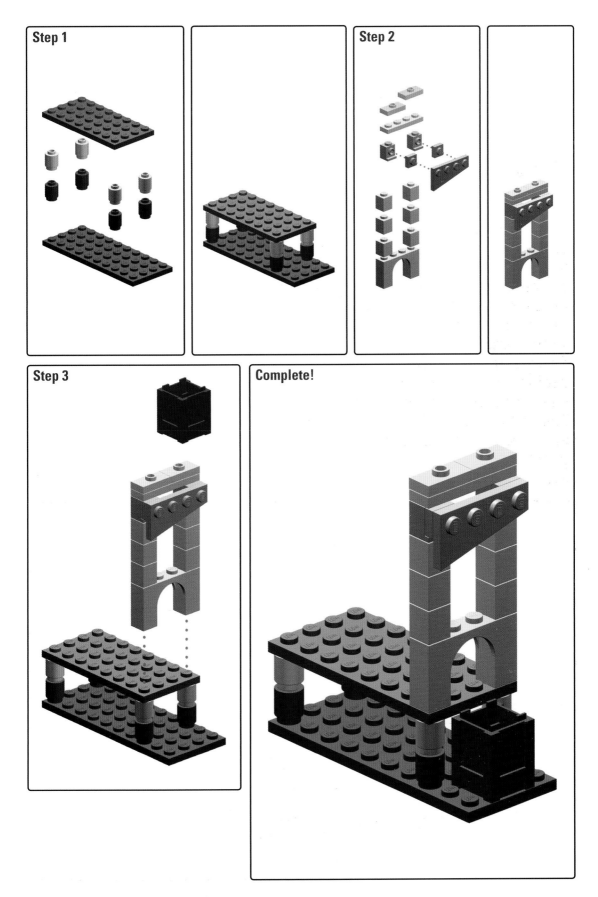

Step 1

Step 2

Step 3

Complete!

RACK
SCALE: 1:44

This macabre piece of machinery, made of a square wooden frame that was equipped with pulley-operated rollers, is a medieval torture device that was used in interrogations. A given victim was unceremoniously strapped to the rack by binding his ankles to one end and his wrists to another. As tension was applied to the set of ropes, the rack would slowly pull apart ligaments, joints, and bones, which caused loud popping noises (and probably screaming—lots of screaming). Because of the excruciating nature of this torture device, some prisoners were even forced to watch this process as a form of intimidation. Using the LEGO figure, both cranks can be used by wrapping LEGO-issue string around the middle post to the lower or upper pulleys and onto the toy occupant.

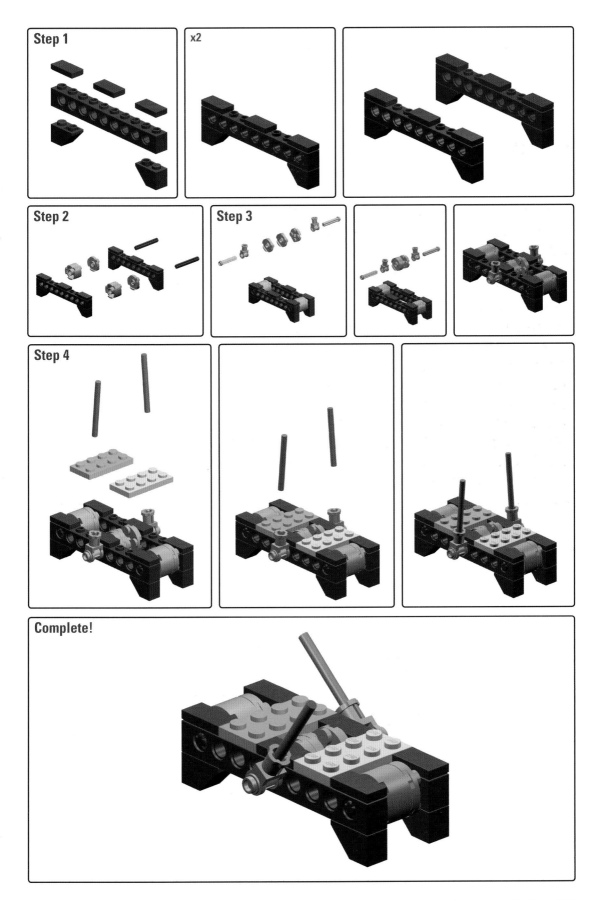

Step 1

x2

Step 2

Step 3

Step 4

Complete!

VIPER COMBAT ROBOT

SCALE: 1:30

Intended as a scout for ground forces in dangerous situations, the VIPeR Combat Robot was released in 2007 and allows troops to have eyes in all places, such as caves and rubble. A robot small enough to fit in a backpack, it can also adapt to changes in terrain, turn itself over, and go up and down stairs. The VIPeR goes in to harm's way to protect soldiers by sniffing out IEDs and detecting enemies and booby traps, but it is also so hardcore that it can be armed to the teeth for intelligent offensive measures.

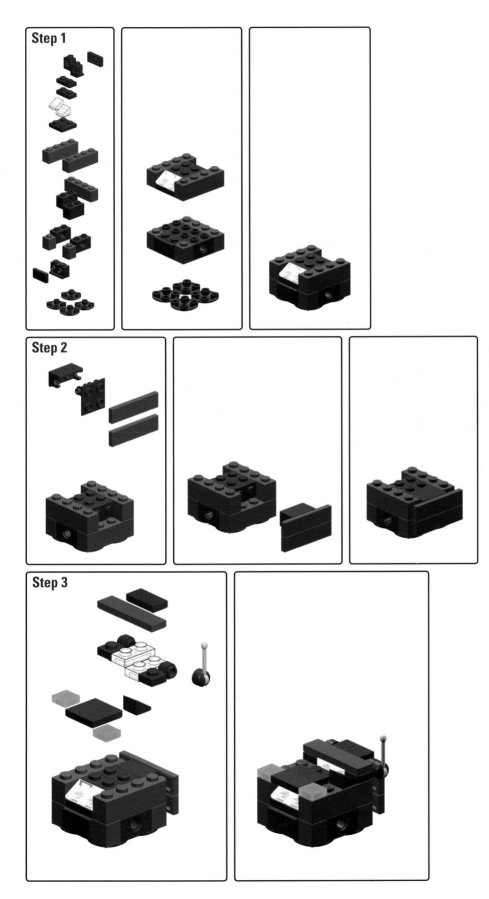

Step 1

Step 2

Step 3

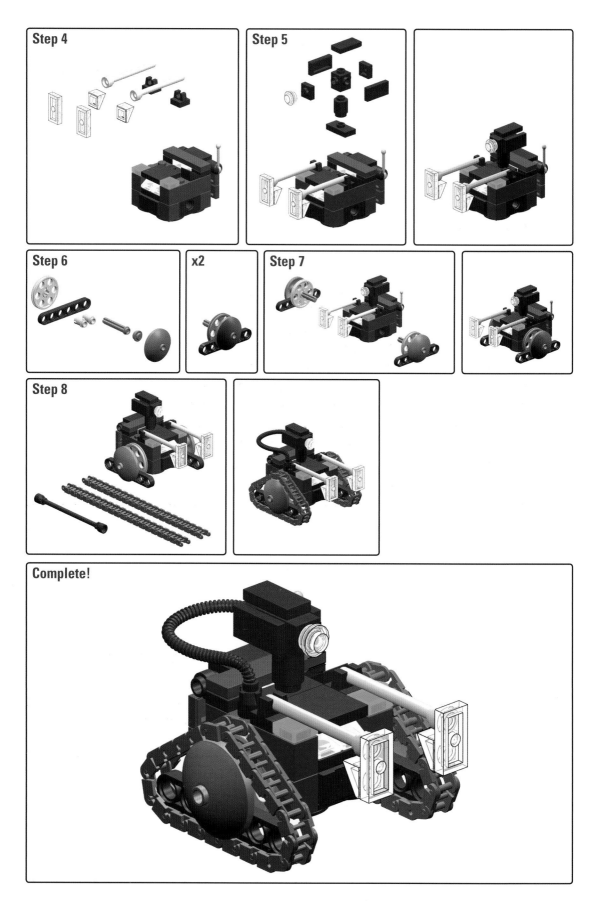

Step 4

Step 5

Step 6

x2

Step 7

Step 8

Complete!

PARIS GUN

SCALE: 1:80

The Paris gun was a long-range artillery weapon used from March to August of 1918 during WWI. It was named such because the Germans used it to hit Paris from three miles away. The weapon was not terribly efficient, and its barrel had to be replaced often; but it was very useful in striking fear into the hearts of Parisians, who did not know when or where the next artillery fire would hit. The Germans destroyed the Paris gun at the end of the war. This LEGO model can be adjusted for tilt, and the entire model rests on a turn table so it can turn 360 degrees.

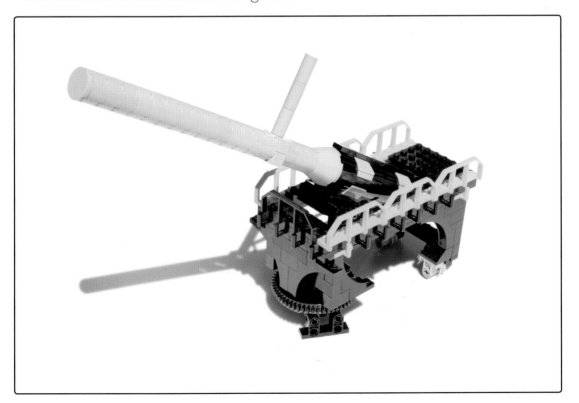

PLATFORM SUPPORTS

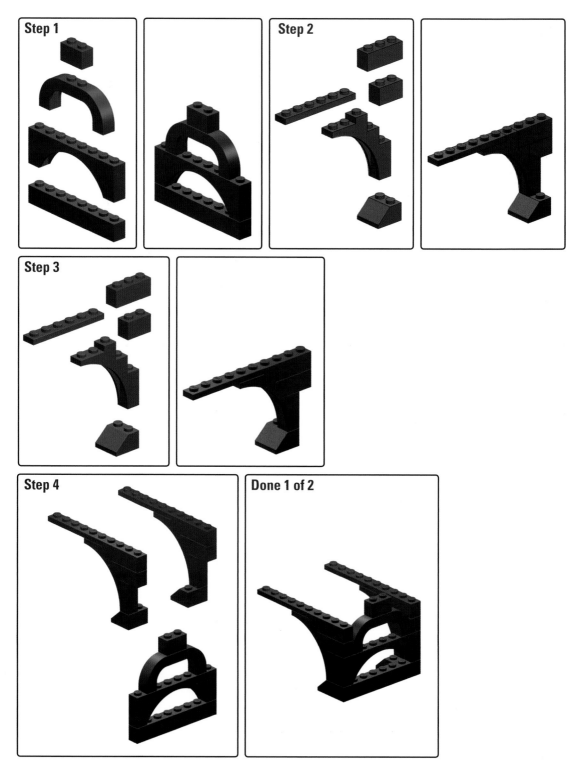

Step 1

Step 2

Step 3

Step 4

Done 1 of 2

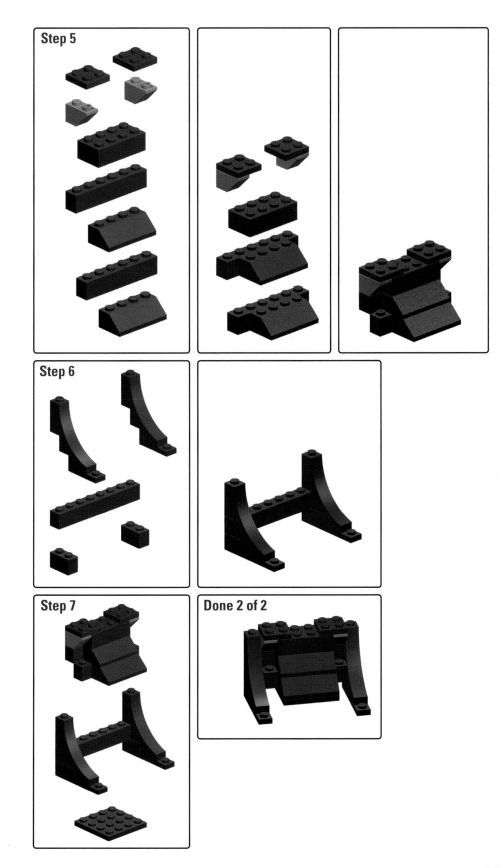

Step 5

Step 6

Step 7

Done 2 of 2

PLATFORM

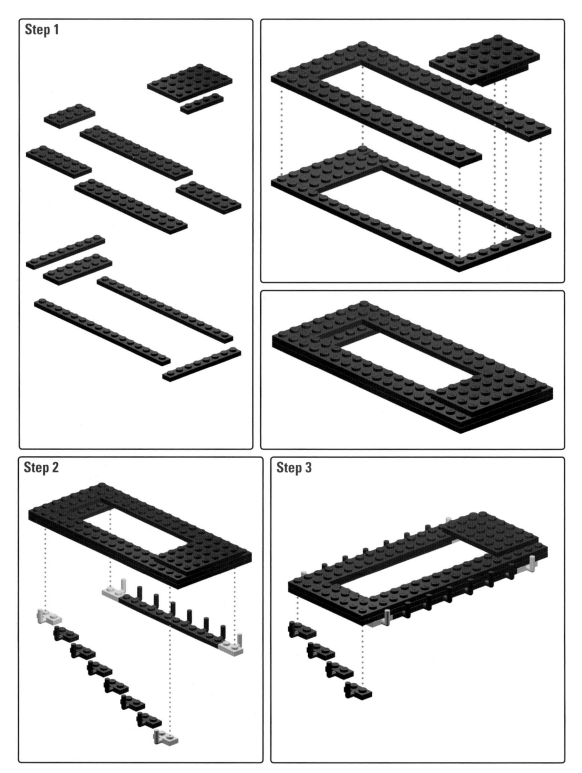

Step 1

Step 2

Step 3

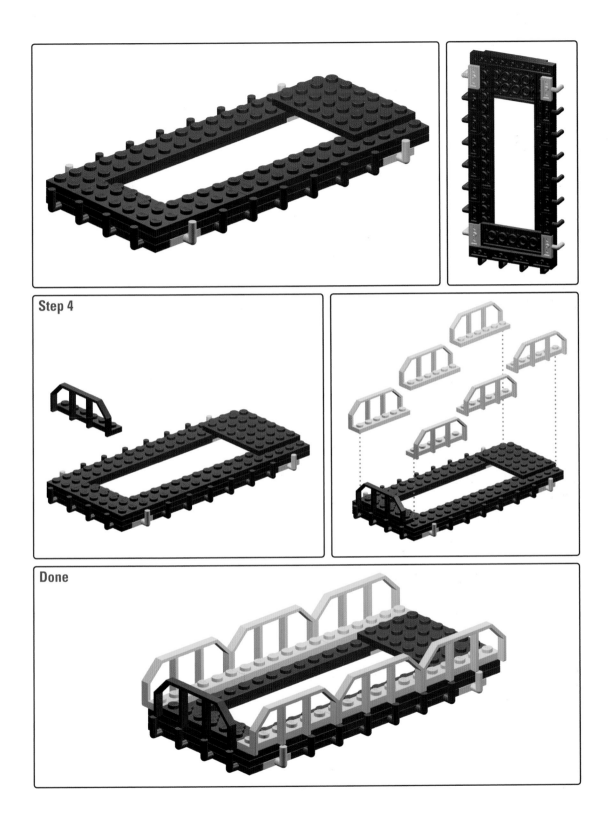

Step 4

Done

Step 1

Step 2

Step 3

Step 4

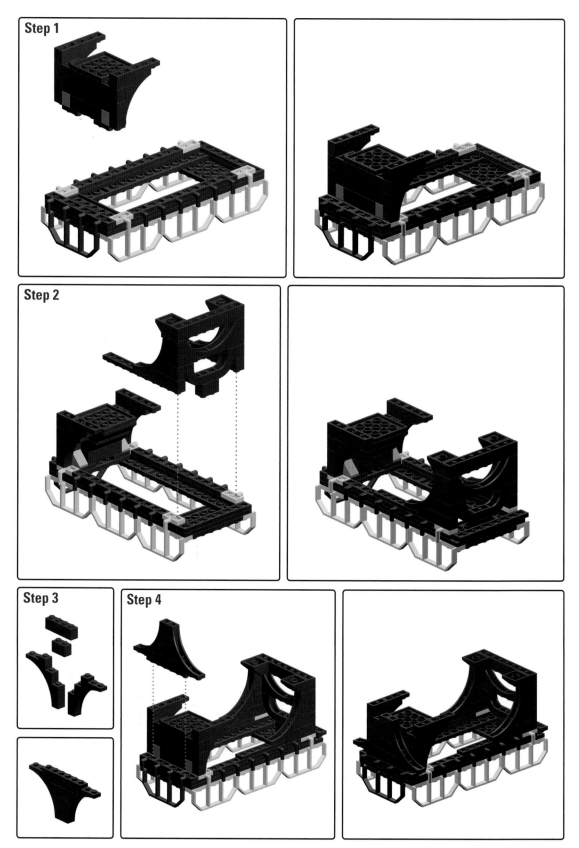

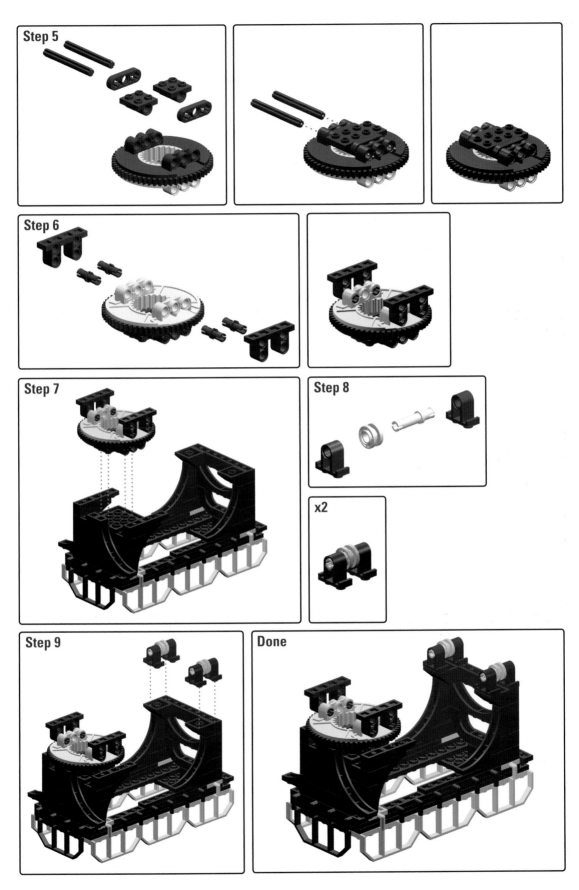

Step 5

Step 6

Step 7

Step 8

x2

Step 9

Done

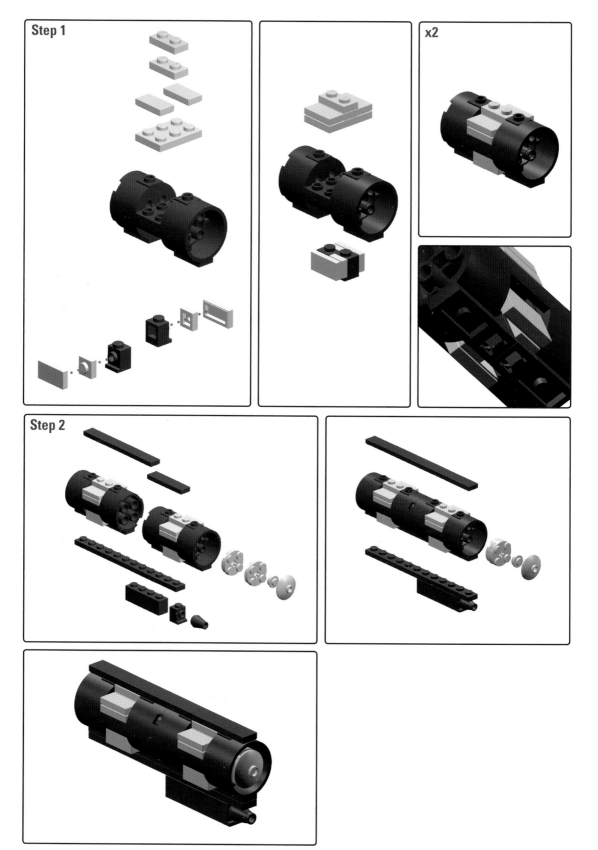

Step 1

x2

Step 2

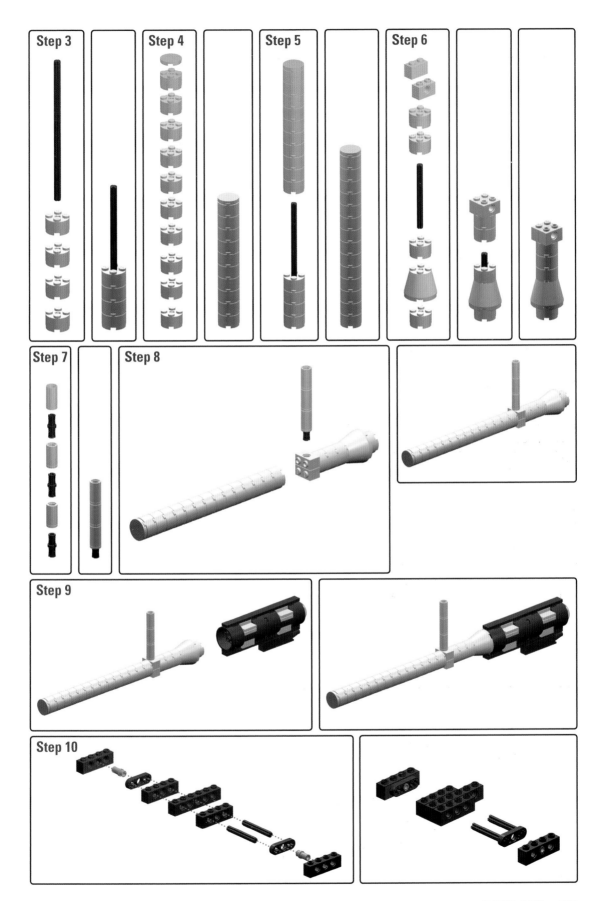

Step 3

Step 4

Step 5

Step 6

Step 7

Step 8

Step 9

Step 10

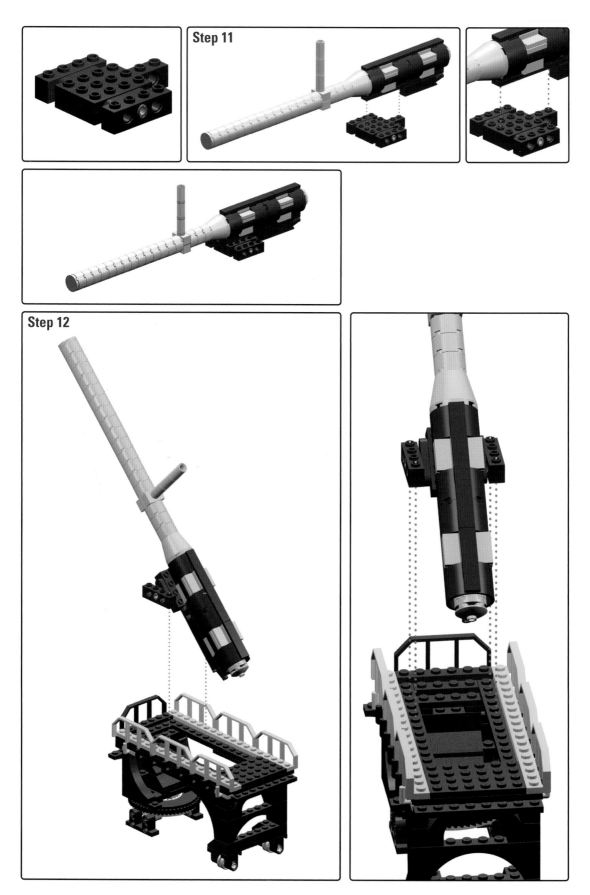

Step 11

Step 12

Complete!

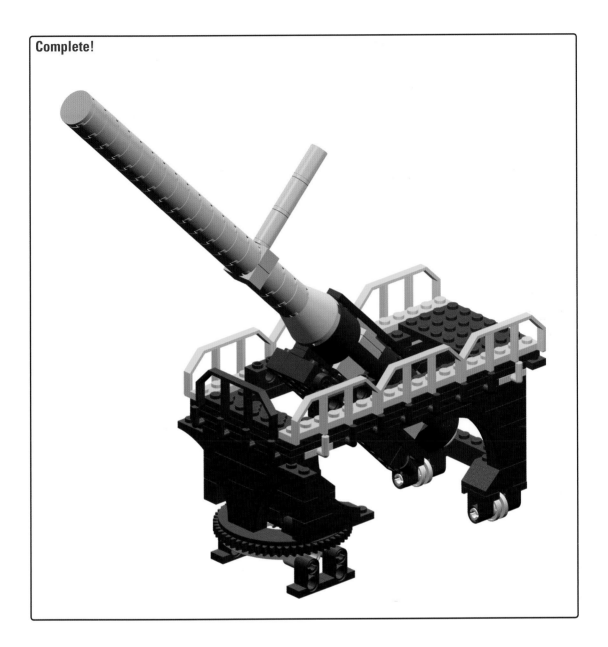

SCYTHED CHARIOT
SCALE: 1:44

Want to make your chariot better? Why not
add swords? That's the idea behind the scythed
chariot, a modified war chariot with serrated
blades or scythes attached to the wheels, which
would cut through infantrymen and the legs of
attacking horses as this vehicle ran through the
battlefield. Persians were documented as using
scythed chariots during the Greco-Persian wars
in 467 BCE, and there is archaeological evidence
that they were used in pre-imperial China as well.

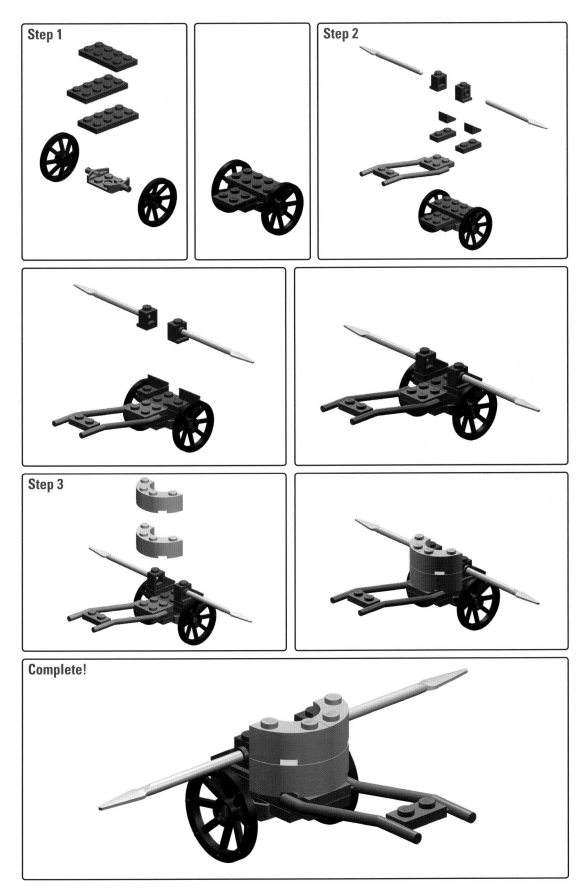

Step 1

Step 2

Step 3

Complete!

KUSARIGAMA
SCALE: 1:1

Not your grandma's nun-chuck. This Japanese weapon has a sickle-shaped end and is attached to a long chain with a heavy weight at the end. The user (or terrifying ninja) swings the weighted chain above his or her head for maximum launching power. Once thrown, the chain disarms the opponent by entangling his or her body or weapon, and the thrower can go in for the kill and attack the opponent with the sickle end. The art of using a *kusarigama* is called *kusarigamajutsu*.

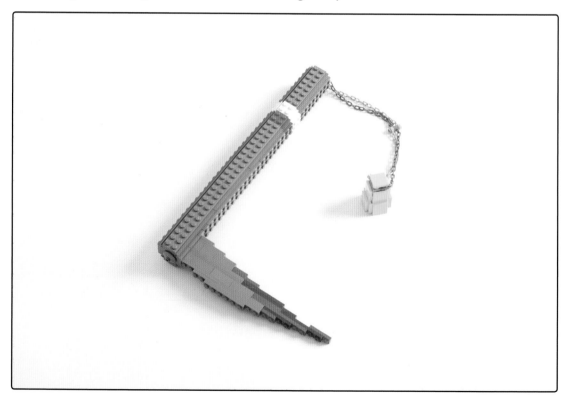

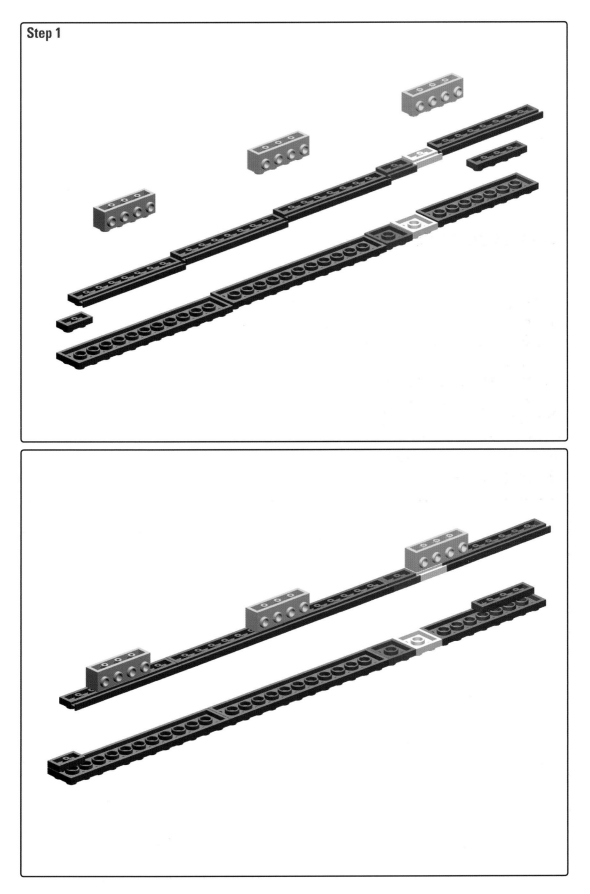

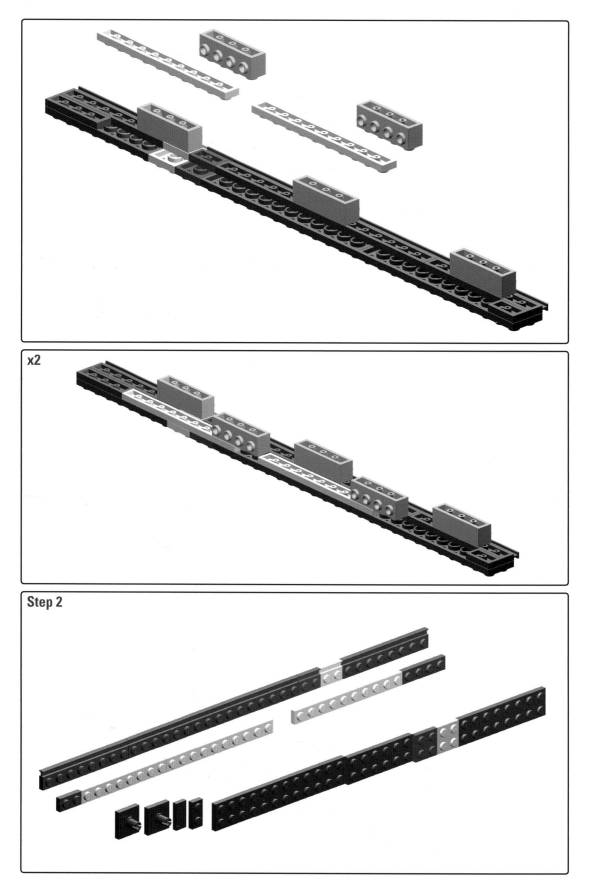

x2

Step 2

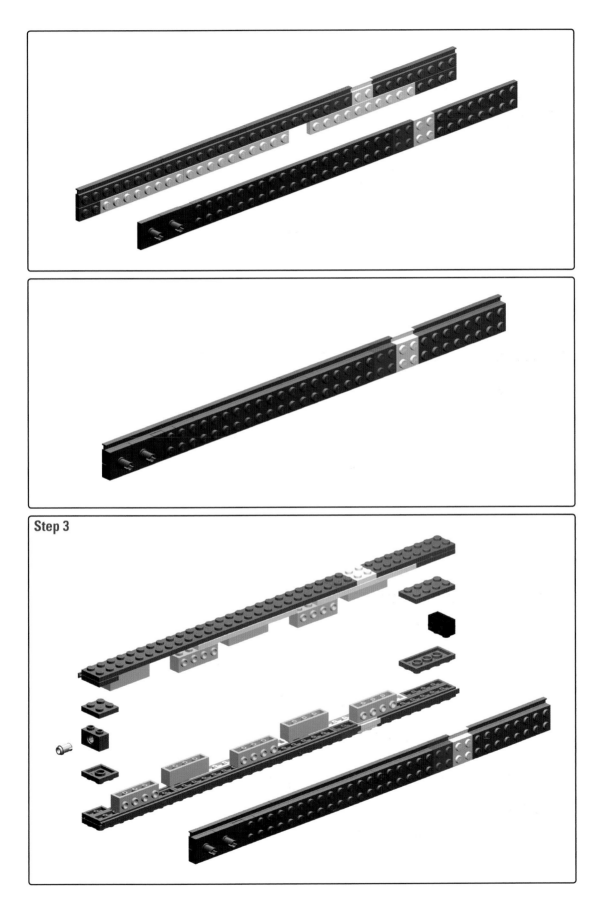

Step 3

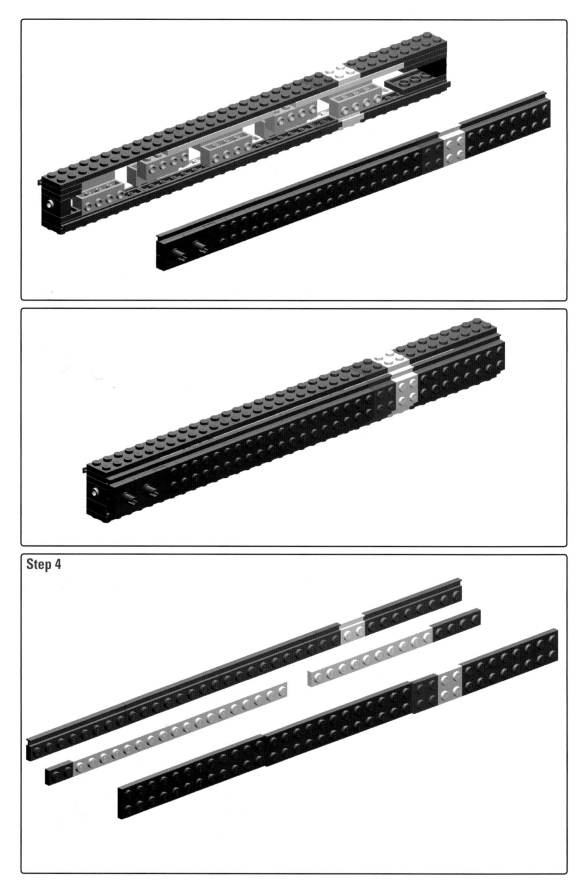

Step 4

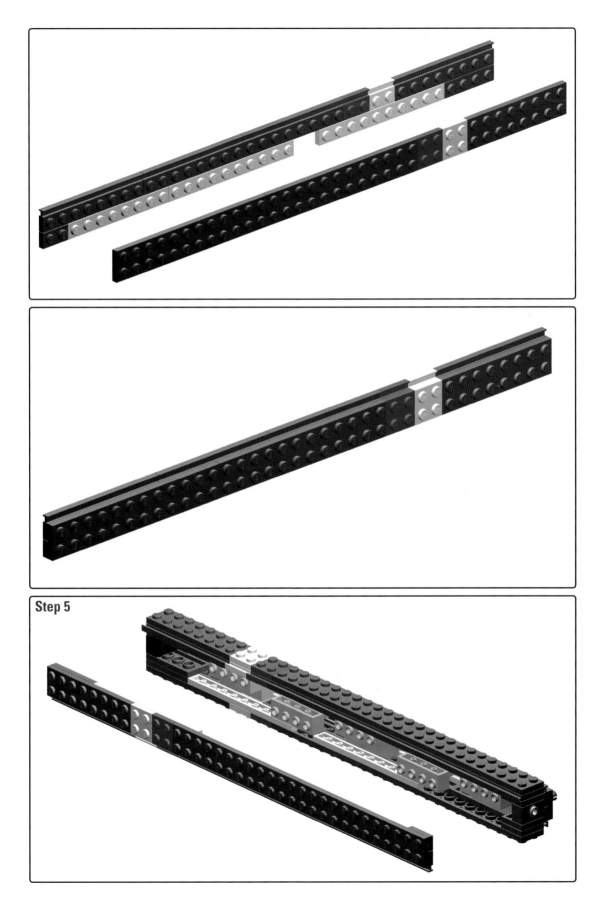

Step 5

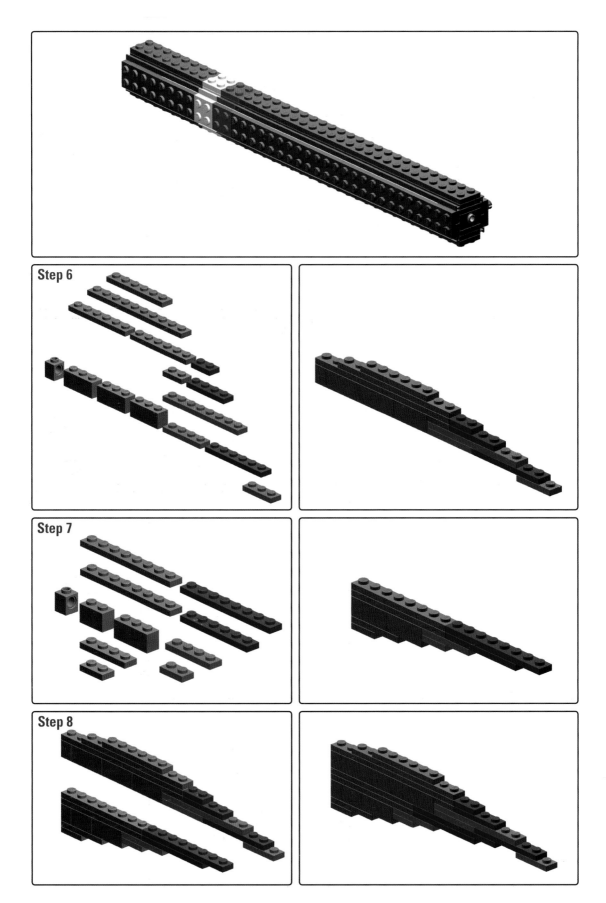

Step 6

Step 7

Step 8

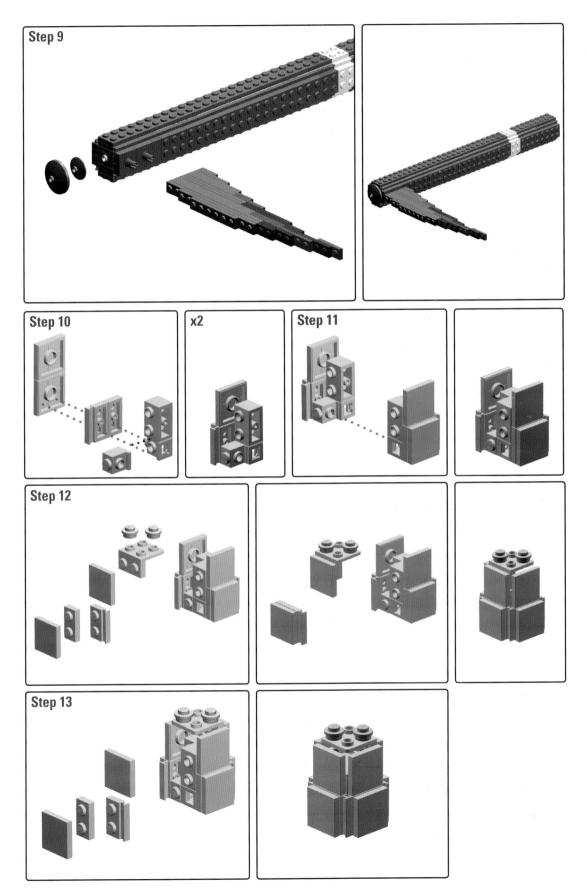

Step 9

Step 10

x2

Step 11

Step 12

Step 13

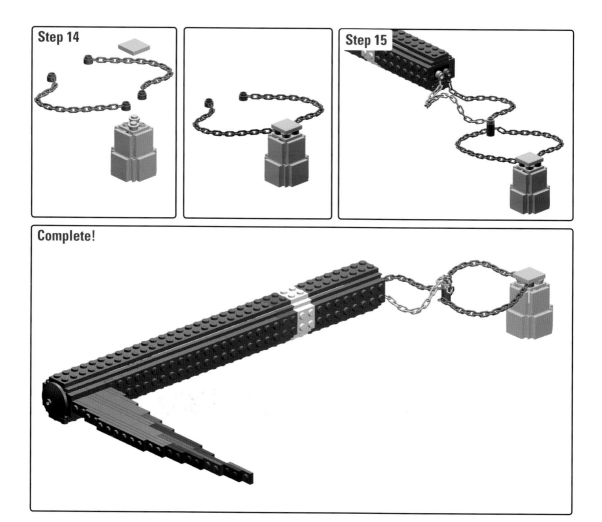

Step 14

Step 15

Complete!

BILL OF MATERIALS

WAR HAMMER

4589	2	3068	1	3795	1	4740	1	30374	1
6143	9	3022	1	6141	2	32028	8	
30361	2	3022	2	4032	3	44728	2	
3023	2	6636	2	4032	1	4274	4	

44

KATANA

6091	4	3068	2	3021	4	3795	4	41770	1
3069	2	3623	2	3021	8	60479	8	2450	2
3023	2	3623	6	3666	2	88646	10	54383	1
3023	2	3022	8	3666	4	3034	8	54384	1
3023	8	3022	4	3020	2	3034	22	4275	4
63864	2	3710	2	3020	4	2445	8	4276	4
2431	2	3710	8	3460	6	41769	1	

158

MAQUAHUITL

3004	10	52107	8	3710	2	3020	2
3622	2	30414	2	6636	2	3795	2
6222	1	3024	12	3021	2	3033	4
47905	4	3023	6	3666	6	6141	4

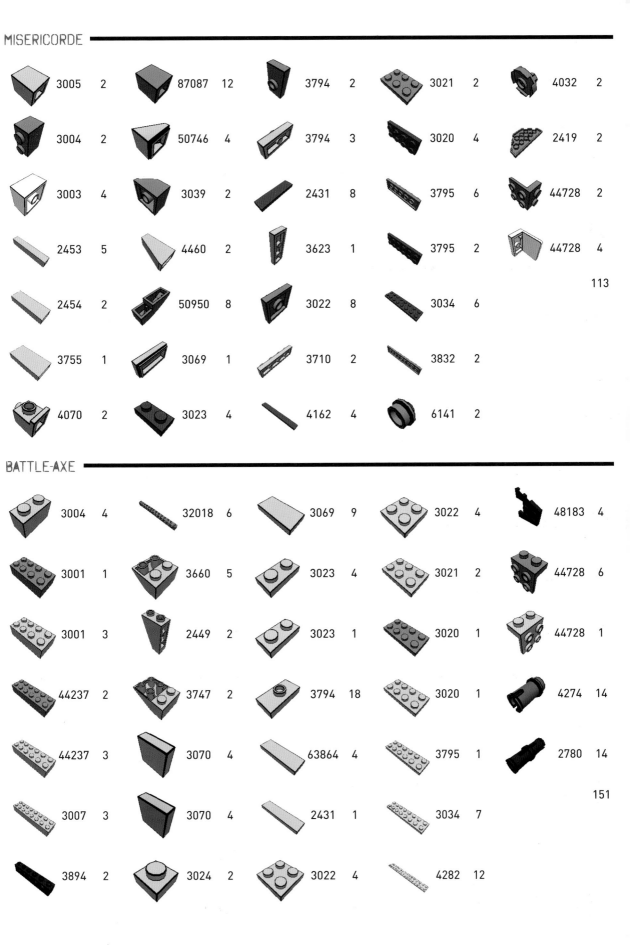

MISERICORDE

3005 2	87087 12	3794 2	3021 2	4032 2
3004 2	50746 4	3794 3	3020 4	2419 2
3003 4	3039 2	2431 8	3795 6	44728 2
2453 5	4460 2	3623 1	3795 2	44728 4
2454 2	50950 8	3022 8	3034 6	113
3755 1	3069 1	3710 2	3832 2	
4070 2	3023 4	4162 4	6141 2	

BATTLE-AXE

3004 4	32018 6	3069 9	3022 4	48183 4
3001 1	3660 5	3023 4	3021 2	44728 6
3001 3	2449 2	3023 1	3020 1	44728 1
44237 2	3747 2	3794 18	3020 1	4274 14
44237 3	3070 4	63864 4	3795 1	2780 14
3007 3	3070 4	2431 1	3034 7	151
3894 2	3024 2	3022 4	4282 12	

CLAYMORE

3005	4	50746	20	3710	2	2445	6	32523	2
3004	2	4286	4	6636	6	3035	20	32525	2
3622	2	3039	8	4162	78	3030	2	32278	2
3003	2	3676	4	3666	4	3029	4	32449	4
3009	1	50950	12	3666	1	3029	4	32017	8
3008	28	3069	20	3020	2	6141	4	4274	24
6222	2	3023	2	3020	2	4150	2	32062	4
4070	16	63864	2	3795	4	41769	4	32002	14
87087	20	2431	6	3795	4	41770	4	32072	2
52107	4	3068	2	3031	2	2419	4		432
30414	42	3022	5	3031	2	43857	1		

HALF-MOON AXE

3004	2	30367	1	3703	2	3069	2	3068	1
3003	1	3942	1	3665	2	3023	1	3068	2
3009	1	87087	6	4287	2	3794	2	3022	3
6143	1	3895	2	61678	4	2431	2	3710	1

HALF-MOON AXE (CONT.)

6636	6	3460	2	6141	12	2780	5
3021	2	3034	14	6141	12	4519	1
3020	1	2445	4	4274	16		114

MORNING STAR

3622	2	3701	1	3039	5	3020	7	4274	16
3003	5	3894	1	3660	5	3795	11	2780	14
4589	25	3702	1	3069	4	3034	4		160
87087	2	2730	1	3023	12	2445	2		
52107	1	3703	4	3022	7	4282	6		
30414	4	3040	4	3710	12	44728	4		

BOWIE KNIFE

3004	1	3008	2	61678	1	3068	4	3666	4
3622	3	87087	8	2339	4	3623	3	3020	2
3003	2	6541	2	3024	1	3022	3	3020	2
3010	2	4286	1	3069	1	3710	1	4477	2
3009	6	4287	1	3794	2	4162	2	3795	2
3001	1	50950	1	63864	2	3021	2	3795	4

BOWIE KNIFE
(CONT.)

Part	Qty	Part	Qty	Part	Qty	Part	Qty	Part	Qty
3795	1	3034	2	3034	2	3032	2	2436	8
60479	1	3034	2	3832	2	3030	2		94

TOMAHAWK

Part	Qty	Part	Qty	Part	Qty	Part	Qty	Part	Qty
3004	1	3702	2	3666	2	3034	6	50304	1
3002	1	3024	4	3020	2	3034	4	50305	1
3001	2	3069	6	3020	2	3032	2	4275	1
3007	2	3023	2	3020	4	3030	2	4276	1
6143	1	3022	1	3795	1	4150	1	44728	4
6233	1	3022	1	3795	1	4032	1	2436	2
4070	8	3710	2	60479	2	41769	2	4274	10
32000	1	3021	1	3031	2	41770	2		99
3701	2	3021	1	3034	2	2450	2		

BLOWGUN

Part	Qty	Part	Qty	Part	Qty	Part	Qty	Part	Qty
3020	4	3832	12	32523	8	32278	12	6558	4
3795	4	32028	8	32316	3	2780	32	6587	16
3034	24	4510	20	32524	8	4519	8	32184	16

3005	4	4287	4	3666	1	41677	2	6558	22
3004	6	4460	4	3460	3	41677	4	6587	8
3622	5	3024	1	44302	2	6632	1	32073	1
2357	1	2431	2	44301	1	32006	1	87083	6
3010	9	3623	2	3482	2	32056	2	42003	8
3002	5	2420	1	32524	2	44374	2	41678	4
3701	2	3710	1	32140	2	32123	1	63869	4
32018	4	6636	2	41239	2	32062	4		

177

3703	4	6636	4	32278	2	2780	4		

3040	3	4162	2	32526	2	6562	3	2449	4
3665	9	3021	1	6629	4	6590	1		

4

3062	4	3023	2	4162	1	6141	3	61252	1
3665	2	3794	2	3460	1	6141	2	2926	1
3023	3	3710	2	3795	2	4740	2	4489	2

CANNON

Part	Qty	Part	Qty	Part	Qty	Part	Qty	Part	Qty
50746	2	3023	1	61252	1	2926	1	30377	2
3070	2	2431	1	63868	1	4489	2		18
3069	1	3666	1	2540	2	4599	1		

TOMMY GUN

Part	Qty	Part	Qty
3062	1	6141	1
50746	1	4085	2
6141	1	92738	1
			7

GATLING GUN

Part	Qty	Part	Qty	Part	Qty
30361	1	61252	1	32123	2
32000	1	63868	1	4274	1
50746	2	2540	2	4519	2
3070	2	50951	1	87994	8
2431	1	2926	1	4599	1
3710	1	4489	2		35
6141	3	6632	2		

WALTHER PPK

Part	Qty	Part	Qty	Part	Qty	Part	Qty	Part	Qty
3005	1	4070	1	50746	1	3678	2	3070	1
3005	1	87087	3	50746	2	6091	4	3024	2
3004	1	87087	17	85984	1	6191	6	3069	1
3004	3	6541	1	3660	1	6005	1	3023	10
3010	1	32000	1	2449	2	6081	1	3023	3

3023	1	2420	1	3666	1	6141	1	60478	1
3794	1	3022	3	3020	6	41769	2	32028	1
3794	5	3710	3	3020	1	41770	2	2452	1
3794	1	3710	2	3460	2	43898	1	30383	1
63864	2	3710	2	4477	2	3933	1	32530	1
2431	1	4162	1	4477	1	3934	1	4274	2
2431	1	3021	7	3795	1	2555	2	6562	1
3068	2	3021	1	2445	2	6019	1	32016	1
3623	7	3666	2	98138	1	4276	1		

149

SIEGE TOWER

3004	16	30136	6	3040	2	3020	3	4489	4
3004	4	30137	15	4460	6	3034	1	2780	1
3004	5	3062	8	3623	2	41539	1	3673	1
30136	4	6541	4	2420	2	2540	1	6020	1
30136	6	50746	4	3710	1	4488	4		

102

TREBUCHET

3003	4	3007	1	3701	2	3666	12	3034	2
3003	1	6541	2	3039	2	3020	2	32525	2
3010	2	3700	4	3039	6	3020	2	32123	4
3010	4	3700	8	3660	2	3020	4	4274	6
3009	2	3700	4	3023	6	3460	4	2780	9
3008	2	32000	2	3068	2	3795	2	3673	2
3008	4	32064	4	3022	2	3795	2	3706	1
44237	9	3701	2	4162	8	3034	8	30374	1

148

BATTERING RAM

2454	4	6222	4	3795	2	4740	1
44237	2	32018	2	3034	6	44728	2
6143	12	2420	2	6141	4	4274	4
30361	1	3666	2	6141	17	6562	4

69

HOWITZER

3005	2	3008	2	6143	3	30361	1	6541	2
3005	2	3062	11	30367	1	47905	2	50746	2

50746	2	3069	2	6636	2	73983	2	3673	2
3040	2	3023	6	3666	2	2444	2	75535	3
3665	2	3023	2	3795	2	4865	1	73587	2
4287	2	3794	2	3031	1	55981	2	61409	2
6182	1	3794	3	6141	1	30391	2	47457	1
3070	2	2420	4	6141	4	4274	2	3938	1
3024	1	3710	1	73983	2	2780	4	3937	1

101

HUMVEE

3005	2	3009	2	3660	4	3069	5	3022	4
2877	6	3008	3	3298	2	3023	3	3022	2
3004	10	4070	6	30499	2	3023	16	3710	5
3065	5	87087	2	3070	1	3794	8	6636	8
3622	2	2921	4	3024	10	3794	6	4162	1
3003	2	32000	4	3024	2	2431	1	3021	2
3010	2	50746	16	2412	1	3068	2	3021	2
3066	1	3040	4	2412	8	3623	2	3666	4

HUMVEE (CONT.)

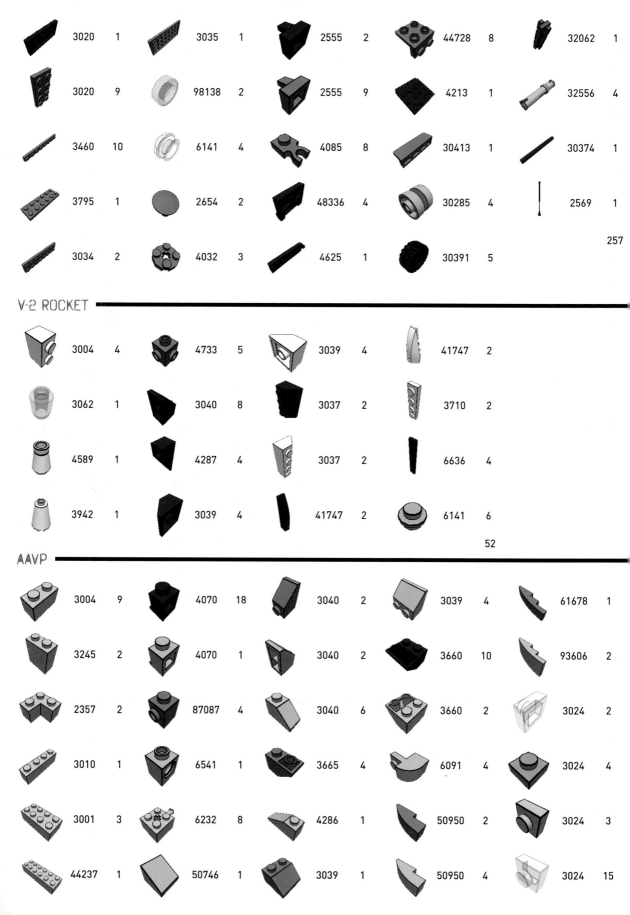

3020	1	3035	1	2555	2	44728	8	32062	1
3020	9	98138	2	2555	9	4213	1	32556	4
3460	10	6141	4	4085	8	30413	1	30374	1
3795	1	2654	2	48336	4	30285	4	2569	1
3034	2	4032	3	4625	1	30391	5		257

V-2 ROCKET

3004	4	4733	5	3039	4	41747	2
3062	1	3040	8	3037	2	3710	2
4589	1	4287	4	3037	2	6636	4
3942	1	3039	4	41747	2	6141	6
							52

AAVP

3004	9	4070	18	3040	2	3039	4	61678	1
3245	2	4070	1	3040	2	3660	10	93606	2
2357	2	87087	4	3040	6	3660	2	3024	2
3010	1	6541	1	3665	4	6091	4	3024	4
3001	3	6232	8	4286	1	50950	2	3024	3
44237	1	50746	1	3039	1	50950	4	3024	15

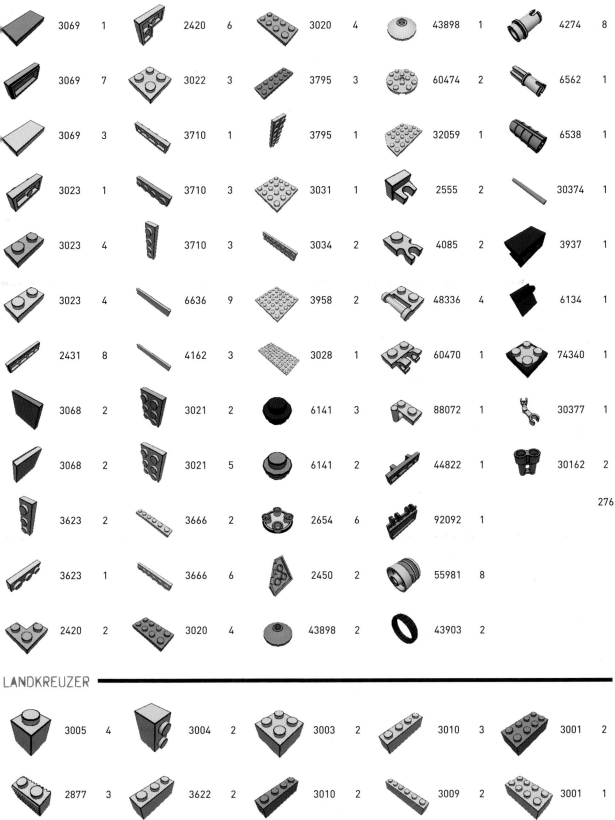

3069	1	2420	6	3020	4	43898	1	4274	8
3069	7	3022	3	3795	3	60474	2	6562	1
3069	3	3710	1	3795	1	32059	1	6538	1
3023	1	3710	3	3031	1	2555	2	30374	1
3023	4	3710	3	3034	2	4085	2	3937	1
3023	4	6636	9	3958	2	48336	4	6134	1
2431	8	4162	3	3028	1	60470	1	74340	1
3068	2	3021	2	6141	3	88072	1	30377	1
3068	2	3021	5	6141	2	44822	1	30162	2
3623	2	3666	2	2654	6	92092	1		276
3623	1	3666	6	2450	2	55981	8	
2420	2	3020	4	43898	2	43903	2	

LANDKREUZER

3005	4	3004	2	3003	2	3010	3	3001	2
2877	3	3622	2	3010	2	3009	2	3001	1

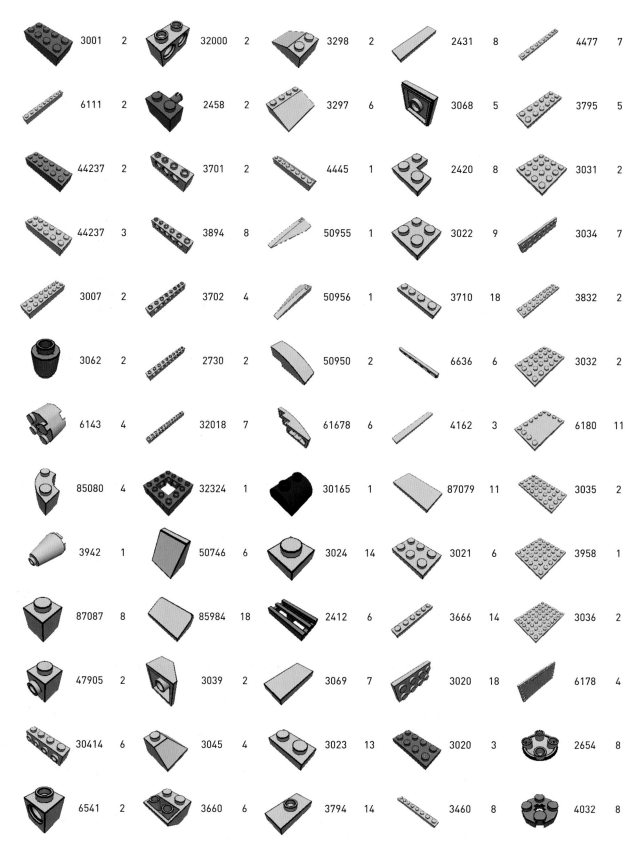

3001	2	32000	2	3298	2	2431	8	4477	7
6111	2	2458	2	3297	6	3068	5	3795	5
44237	2	3701	2	4445	1	2420	8	3031	2
44237	3	3894	8	50955	1	3022	9	3034	7
3007	2	3702	4	50956	1	3710	18	3832	2
3062	2	2730	2	50950	2	6636	6	3032	2
6143	4	32018	7	61678	6	4162	3	6180	11
85080	4	32324	1	30165	1	87079	11	3035	2
3942	1	50746	6	3024	14	3021	6	3958	
87087	8	85984	18	2412	6	3666	14	3036	2
47905	2	3039	2	3069	7	3020	18	6178	4
30414	6	3045	4	3023	13	3020	3	2654	8
6541	2	3660	6	3794	14	3460	8	4032	8

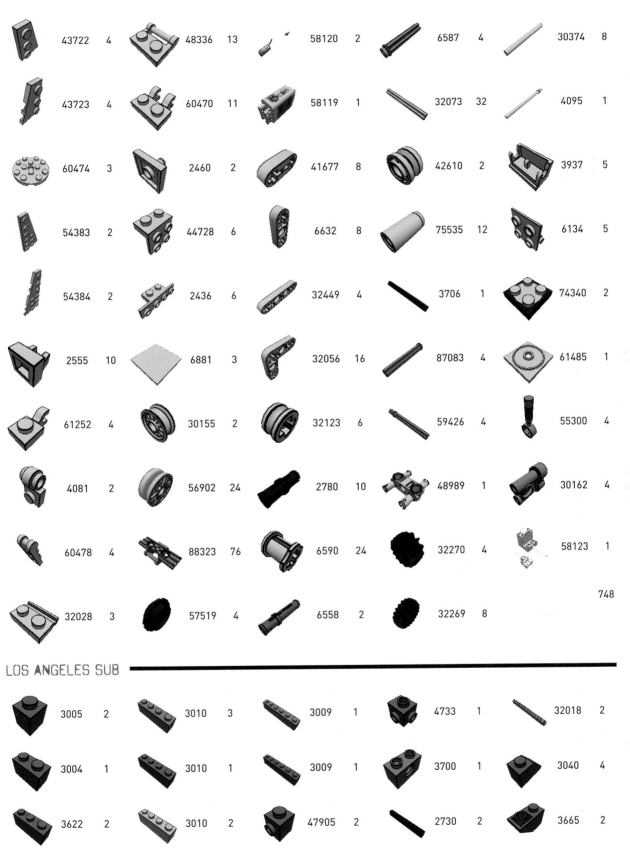

43722	4	48336	13	58120	2	6587	4	30374	8
43723	4	60470	11	58119	1	32073	32	4095	1
60474	3	2460	2	41677	8	42610	2	3937	5
54383	2	44728	6	6632	8	75535	12	6134	5
54384	2	2436	6	32449	4	3706	1	74340	2
2555	10	6881	3	32056	16	87083	4	61485	1
61252	4	30155	2	32123	6	59426	4	55300	4
4081	2	56902	24	2780	10	48989	1	30162	4
60478	4	88323	76	6590	24	32270	4	58123	1
32028	3	57519	4	6558	2	32269	8		748

LOS ANGELES SUB

3005	2	3010	3	3009	1	4733	1	32018	2
3004	1	3010	1	3009	1	3700	1	3040	4
3622	2	3010	2	47905	2	2730	2	3665	2

LOS ANGELES SUB (CONT.)

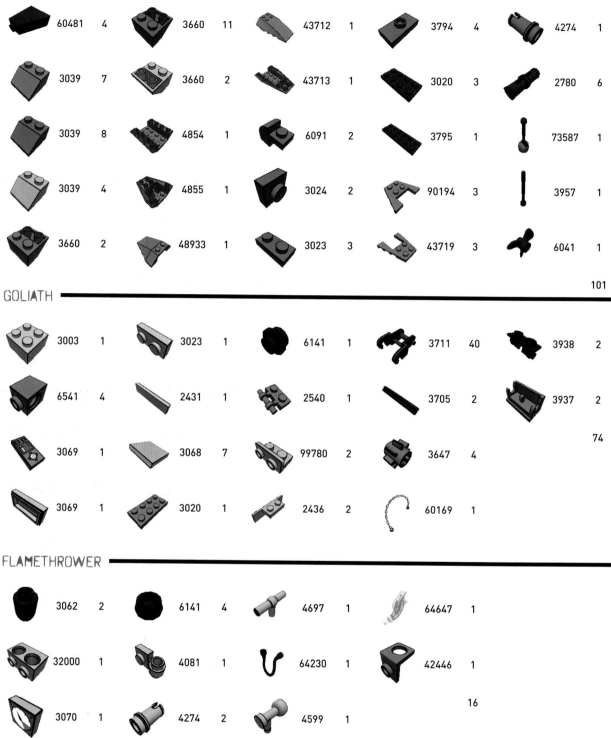

Part	Qty	Part	Qty	Part	Qty	Part	Qty	Part	Qty
60481	4	3660	11	43712	1	3794	4	4274	1
3039	7	3660	2	43713	1	3020	3	2780	6
3039	8	4854	1	6091	2	3795	1	73587	1
3039	4	4855	1	3024	2	90194	3	3957	1
3660	2	48933	1	3023	3	43719	3	6041	1

101

GOLIATH

Part	Qty	Part	Qty	Part	Qty	Part	Qty	Part	Qty
3003	1	3023	1	6141	1	3711	40	3938	2
6541	4	2431	1	2540	1	3705	2	3937	2
3069	1	3068	7	99780	2	3647	4		
3069	1	3020	1	2436	2	60169	1		

74

FLAMETHROWER

Part	Qty	Part	Qty	Part	Qty	Part	Qty
3062	2	6141	4	4697	1	64647	1
32000	1	4081	1	64230	1	42446	1
3070	1	4274	2	4599	1		

16

GUILLOTINE

3005	6	4070	2	3794	2	3030	1
3062	4	6182	1	3710	1	41769	1
3062	4	3024	2	3035	1	61780	1

26

RACK

6143	2	3069	6	4032	5	32039	2
2730	2	3020	1	3705	2	30374	2
3665	4	3020	1	6587	2		

29

VIPER

3005	2	50746	1	3794	1	43898	2	6587	2
3010	3	50746	4	2431	3	2555	2	64230	1
3062	1	3070	2	3068	1	4081	2	73587	1
4070	4	3024	2	3022	1	3711	38	3957	2
4733	1	3069	5	6141	2	4185	2	3937	1
3700	2	3023	3	6141	1	32063	2	6134	1
32064	2	3023	4	2654	4	4274	4		

109

PARIS GUN

3004 6	3701 4	3022 2	6141 1	32556 2
3004 1	3894 1	3710 1	4150 1	32073 2
3622 2	85984 8	4162 1	4032 2	42610 2
3010 2	3665 2	3021 2	4740 1	75535 3
3009 2	3039 2	3666 4	32529 4	3706 1
3001 1	3037 2	3020 1	32530 4	3708 1
3008 2	6183 1	3460 2	88072 16	50163 1
4589 1	3308 1	4477 2	88072 4	6583 1
6143 18	2339 4	3795 3	2817 2	6583 6
6233 1	30099 2	60479 1	6632 4	172
30360 2	3024 4	3031 1	32002 2	
4070 5	3023 4	2445 2	2780 7	
3700 1	2431 1	3032 1	3705 2	

SCYTHED CHARIOT

85080 4	50746 2	3022 1	6157 1	4497 2
4070 2	3023 2	3020 3	2470 2	19

3004	1	3023	7	3710	5	3460	1	32028	4
3622	4	3023	1	3710	1	4477	1	32028	4
4070	4	3023	1	3710	2	4477	8	32028	4
52107	1	3023	4	3666	3	3795	1	2460	2
52107	2	3068	9	3666	2	3034	4	44728	1
30414	10	3623	1	3020	2	2445	7	4510	16
6541	2	3022	2	3460	2	6141	2	4274	1
3700	1	3022	4	3460	2	4740	1	60169	3
3069	1	3022	4	3460	1	43898	1	60169	3

143

All part numbers are exported from LEGO Digital Designer (v4.3),
a free program available from the LEGO Group. If you cannot locate or
identify a part, please search part numbers from within that program.

ALSO AVAILABLE

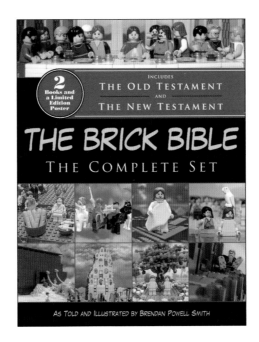

The Brick Bible: The Complete Set

by Brendan Powell Smith

The Brick Bible series has taken the world by storm, and now, for the first time, Brendan Powell Smith's visually striking *The Brick Bible: A New Spin on the Old Testament* and *The Brick Bible: The New Testament* are available in a beautiful hardcover box set. With over two thousand color photographs depicting the major narrative scenes of the Bible, this slipcovered set (including new material and a bonus two-sided, full-color poster) is the gift you've been wanting to give your LEGO-loving friends and religious family members for holidays, birthdays, or just because.

Book one includes scenes from the Old Testament—the creation of the world, the temptation of Adam and Eve, the great flood, the presentation of the ten commandments to Moses on Mt. Sinai, and more. The Old Testament is a complex text, but Smith's "brick" illustrations help bring out the nuances of each scene and will make you reconsider the way you look at the Bible—and LEGO.

Book two offers a new spin on the story of Jesus. Smith portrays Jesus's birth, miracles, last supper, and death and resurrection with meticulous attention to detail. From the fate of Judas to the life of Paul and his letters to the Ephesians; from the first book burning to the book of Revelations, this is the New Testament as you've never experienced it before. Find a place on your shelf for this beautiful two-book collector's set and continue to be amazed at the Bible as illustrated by your favorite toy.

$29.95 Hardcover · ISBN 978-1-62636-177-5

ALSO AVAILABLE

A Million Little Bricks

The Unofficial Illustrated History of the LEGO Phenomenon

by Sarah Herman

There aren't many titles that haven't been bestowed on LEGO toys, and it's not hard to see why. From its inception in the early 1930s right up until today, the LEGO Group's history is as colorful as the toys it makes. Few other playthings share the LEGO brand's creative spirit, educative benefits, resilience, quality, and universal appeal. The LEGO name is now synonymous with playtime, but it wasn't always so. This history charts the birth of the LEGO Group in the workshop of a Danish carpenter and its steady growth as a small, family-run toy manufacturer to its current position as a market-leading, award-winning brand. The company's ever-increasing catalog of products—including the earliest wooden toys, plastic bricks, play themes, and other building systems such as DUPLO, Technic, and MINDSTORMS—are chronicled in detail, alongside the manufacturing process, LEGOLAND parks, licensed toys, and computer and video games.

Learn all about how LEGO pulled itself out of an economic crisis and embraced technology to make building blocks relevant to twenty-first-century children, and discover the vibrant fan community of kids and adults whose conventions, websites, and artwork keep the LEGO spirit alive. As nostalgic as it is contemporary, *A Million Little Bricks* will have you reminiscing about old Classic Space sets, rummaging through the attic for forgotten Minifigure friends, and playing with whatever LEGO bricks you can get your hands on (even if it means sharing with your kids).

$16.95 Paperback · ISBN 978-1-62636-118-8

ALSO AVAILABLE

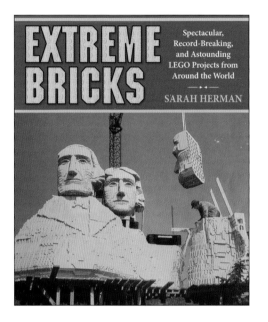

Extreme Bricks

Spectacular, Record-Breaking, and Astounding LEGO Projects from around the World

by Sarah Herman

From the Great Pyramid of Giza and Stonehenge to the Colosseum and the Taj Mahal, man has never shied away from an extreme building challenge, and the LEGO builders of the twenty-first century are no different. Whether they're re-creating the works of ancient masters in brick form, building life-size superheroes, breaking world records with skyscraping towers, or firmly adjusting their thinking caps to program plastic robots, some LEGO fans are taking their passion for plastic to the extreme.

Sarah Herman, the author of *A Million Little Bricks: The Unofficial Illustrated History of the LEGO Phenomenon*, has brought together some of the world's most ambitious builders in a fact-filled showcase of truly mind-blowing models for this exciting illustrated book. Extreme Bricks chronicles the first attempts at large-scale models embarked on by the LEGO Group as well as the early work of LEGOLAND artists and builders and contemporary LEGO Certified Professionals who build big for a living. It charts the rise of AFOLs (adult fans of LEGO) and their increasingly spectacular models and gargantuan collections. Packed with two hundred color photographs to shock and surprise, the book also explores the popular building competitions and includes a closer look at the shockingly smart LEGO MINDSTORMS robots that can do everything from solving a Rubik's Cube to building their own LEGO models.

$24.95 Hardcover • ISBN 978-1-62636-212-3